Financial Accounting
財務會計(雙語版)

主　　編○秦戈雯、池昭梅
副主編○劉德宏、朱　懿
　　　　吳　丹、陽春暉

財經錢線

前　言

　　會計的國際化進程是會計發展的新常態，是一種必然趨勢，全世界的會計準則制定機構都走上了會計準則國際趨同和等效之路。高校是培育國際化會計人才的根據地，越來越多的高校開設了會計專業雙語課程，以培養具有國際視野、通曉中國會計準則及國際會計準則的高素質會計專業人才。在促進國際會計專業建設進程中，在促進會計專業核心課程教學改革發展進程中，優秀教材建設對於培養會計人才的重要性是顯而易見的。財務會計雙語課程是普通高等教育國際會計專業的基礎課程，也是該專業的核心課程。該門課程的學習效果直接影響學生的學習興趣及後續專業課程的學習。

　　《財務會計（雙語版）》在編寫的過程中突出了以下特點：第一，本書依據最新的國際會計準則，各項經濟業務的處理參照國際會計慣例。第二，以會計的基本理論為基礎，闡述了會計的基本原則，並通過舉例清晰地解釋了各項經濟業務的處理方法。第三，本書以英語為主，其中專業知識的重難點有適當的中文解釋，有助於學生在學習國際會計時更好地在英文的環境下學習和理解相關專業理論與方法，增強學生的學習興趣。本書適用於國際會計專業本科生的教學，同時也可供會計人員培訓和自學使用，是學習國際會計準則的入門教材。

　　本書的編著歷時一年多，數易其稿，反覆修改，力求做到深入淺出，通俗易懂。由於我們編寫水平有限，書中有不足之處在所難免，懇請廣大讀者和同行們提出寶貴意見和建議，以臻完善。

編　者

Contents

Chapter 1	**Introduction to Accounting**	1)
Unit 1	Overview of Accounting	1)
Unit 2	Types of Business	2)
Unit 3	The Users of Financial Information	5)
Chapter 2	**The Qualitative Characteristics of Financial Information**	8)
Unit 1	Fundamental Accounting Assumptions	8)
Unit 2	The Qualitative Characteristics of Financial Information	9)
Chapter 3	**Source Document, Records and Books of Prime Entry**	14)
Unit 1	Types of Source Documents	15)
Unit 2	Books of Prime Entry	17)
Chapter 4	**Ledger Accounts & Control Account**	22)
Unit 1	General Ledger	22)
Unit 2	Control Accounts	24)
Chapter 5	**Double Entry Bookkeeping**	32)
Unit 1	Accounting Elements and Accounting Equation	32)
Unit 2	Introduction of Double Entry Bookkeeping	36)
Unit 3	The Use of Double Entry Bookkeeping	38)
Chapter 6	**Sales**	44)
Unit 1	Sales Revenue	44)
Unit 2	Sales Tax	45)
Chapter 7	**Inventory**	50)
Unit 1	Accounting for Inventory	50)
Unit 2	Counting Inventory	54)
Unit 3	Valuing Inventory	54)

Chapter 8 Non-current Assets 63)
Unit 1 Acquisition of a Non-current Asset 63)
Unit 2 Depreciation 66)
Unit 3 Disposal of Non-current Asset 70)
Unit 4 Revaluation of Non-current Assets 73)

Chapter 9 Intangible Non-current Assets 77)
Unit 1 Intangible Assets 77)
Unit 2 Research and Development Costs 78)

Chapter 10 Accruals and Prepayments 83)
Unit 1 Accrued Expenditure and Prepaid Expenditure 83)
Unit 2 Accrued Income and Prepaid Income 86)

Chapter 11 Provision and Contingencies 89)
Unit 1 Provision 89)
Unit 2 Contingencies 91)

Chapter 12 Bank Reconciliation 93)
Unit 1 Bank Statement and Cash Book 93)
Unit 2 Bank Reconciliation 94)

Chapter 13 Correction of Errors 97)
Unit 1 Errors Identifiable in Trial Balance 97)
Unit 2 Suspense Account 102)
Unit 3 Error Not Identifiable in Trial Balance 104)

Chapter 14 Preparing a Trial Balance 110)
Unit 1 Trial Balance 110)
Unit 2 Step of Preparing a Trial Balance 112)

Chapter 15 Prepare Basic Financial Statements 116)
Unit 1 Statement of Financial Position 116)
Unit 2 Statement of Profit or Loss 119)
Unit 3 Preparing Financial Statements 121)

目　錄

第一章　會計概述　　　　　　　　　　　　　　　　　　　　（1）
　　第一節　會計概論　　　　　　　　　　　　　　　　　　（1）
　　第二節　企業的類別　　　　　　　　　　　　　　　　　（2）
　　第三節　財務信息的使用者　　　　　　　　　　　　　　（5）

第二章　財務信息的質量特徵　　　　　　　　　　　　　　　（8）
　　第一節　會計基礎假設　　　　　　　　　　　　　　　　（8）
　　第二節　財務信息的質量特徵　　　　　　　　　　　　　（9）

第三章　原始憑證、會計記錄與原始記錄簿　　　　　　　　（14）
　　第一節　原始憑證的種類　　　　　　　　　　　　　　（15）
　　第二節　原始記錄簿　　　　　　　　　　　　　　　　（17）

第四章　分類帳戶和控制帳戶　　　　　　　　　　　　　　（22）
　　第一節　分類帳戶　　　　　　　　　　　　　　　　　（22）
　　第二節　控制帳戶　　　　　　　　　　　　　　　　　（24）

第五章　復式記帳法　　　　　　　　　　　　　　　　　　（32）
　　第一節　會計要素和會計等式　　　　　　　　　　　　（32）
　　第二節　復式記帳法概述　　　　　　　　　　　　　　（36）
　　第三節　復式記帳法的使用　　　　　　　　　　　　　（38）

第六章　銷售　　　　　　　　　　　　　　　　　　　　　（44）
　　第一節　銷售收入　　　　　　　　　　　　　　　　　（44）
　　第二節　銷售稅　　　　　　　　　　　　　　　　　　（45）

第七章　存貨　　　　　　　　　　　　　　　　　　　　　（50）
　　第一節　存貨的會計處理　　　　　　　　　　　　　　（50）
　　第二節　存貨的盤點　　　　　　　　　　　　　　　　（54）
　　第三節　存貨的計價　　　　　　　　　　　　　　　　（54）

第八章　非流動資產　(63)
　　第一節　購置非流動資產　(63)
　　第二節　折舊　(66)
　　第三節　非流動資產的處置　(70)
　　第四節　非流動資產重估　(73)

第九章　無形非流動資產　(77)
　　第一節　無形資產　(77)
　　第二節　研發支出　(78)

第十章　應計項目和預付項目　(83)
　　第一節　應計費用和預付費用　(83)
　　第二節　應計收入和預收收入　(86)

第十一章　準備及或有事項　(89)
　　第一節　準備　(89)
　　第二節　或有事項　(91)

第十二章　銀行存款餘額調節　(93)
　　第一節　銀行對帳單和銀行存款日記帳　(93)
　　第二節　銀行存款餘額調節　(94)

第十三章　差錯更正　(97)
　　第一節　試算平衡表能發現的差錯　(97)
　　第二節　臨時帳戶　(102)
　　第三節　試算平衡表中不能發現的差錯　(104)

第十四章　編製試算平衡表　(110)
　　第一節　試算平衡表　(110)
　　第二節　試算平衡表的編製步驟　(112)

第十五章　編製財務報表　(116)
　　第一節　資產負債表　(116)
　　第二節　利潤表　(119)
　　第三節　財務報表的編製　(121)

Chapter 1　Introduction to Accounting

Accounting is vital to a strong company, keeping track of the business' finances and its continued profitability. Without accounting, a business owner would not know how much money was coming in or going out, or how to plan for the future. The actions taken by accounting professionals – from bookkeepers to certified public accountants (CPAs) – make it possible to monitor the company's financial status and provide reports and projections that affect the organization's decisions.

Unit 1　Overview of Accounting

1.1　The Purpose of Financial Reporting

The accounting system of a business records and summarises the financial performance/position of a business over/at a certain period of time. This information is crucial to various stakeholders of the business, who will analyse that information to make significant economic decisions. It is vital that these stakeholders have good quality information to be able to make good quality decisions.

・經濟信息記錄在原始記錄簿。

・This information is recorded in books of prime entry.

・原始記錄簿中的信息匯總分析後過入到總分類帳戶中。

・This information is then analysed in the books of prime entry and the totals are posted to the ledger accounts.

・最後匯總信息並編製財務報表。

・This information is summarised in the financial statements.

1.2 Financial Accounting and Management Accounting

1.2.1 Financial Accounting

財務會計是企業報告經營成果和財務狀況的一種方法。

Financial accounting is mainly a method of reporting the results and financial position of a business. It is not primarily concerned with providing information towards the more efficient running of the business. Although financial accounts are of interest to management, their principal function is to satisfy the information needs of persons not involved in running the business. They provide historical information.

財務會計提供歷史信息。

1.2.2 Management Accounting

管理會計是企業管理信息系統，為企業的管理活動分析並提供數據信息。

Management (or cost) accounting is a management information system which analyses data to provide information as a basis for managerial action. The concern of a management accountant is to present accounting information in the form most helpful to management.

The information needs of management go far beyond those of other account users. Managers have the responsibility of planning and controlling the resources of the business. Therefore they need much more detailed information. Thus management accounting is an integral part of management activity concerned with identifying, presenting and interpreting information used for:

- 制定戰略；
- 計劃並控制經濟活動；
- 決策；
- 充分利用經濟資源。

- formulating strategy;
- planning and controlling activities;
- making decision;
- optimising the use of resources.

Unit 2 Types of Business

Businesses range enormously in size, from large oil companies, mobile telephone operators and big supermarket chains down to small one-person operations, such as a plumber or a window cleaner. They all have something in common. They sell goods or services for money.

Most types of businesses can be classified in terms of who is involved in them and how the organisations operate.

2.1 Sole Trader/Proprietorship

個體經營、獨資企業

An individual can run his or her own business, either alone as a one-person operation or as a business owner with several employees. The owner is entitled to all the profits (that is what is left after expenses have been deducted from income) and suffers all the losses made by the business.

所有者享有企業所有的利潤並承擔所有的損失。

A sole trader may be involved in following items.
· Manufacturing: making something from raw materials, for example, furniture.
· Trading: buying and selling goods as in a small shop.
· Service: providing a service for customers such as hairdressing.

Accounting conventions recognise the business as a separate entity from its owner. However, legally, the business and personal affairs of a sole trader are not distinguished in any way. The most important consequences of this is that a sole trader has complete personal unlimited liability. Business debts which cannot be paid from business assets must be met from sale of personal assets, such as a house or a car.

在法律上，獨資企業的經營業務和個人事務是不區分的。

最重要的結果就是獨資企業所有人承擔完全個人無限責任。

The advantages of operating as a sole trader include flexibility and autonomy. A sole trader can manage the business as he or she likes and can introduce or withdraw capital at any time.

2.2 Partnership

合夥企業

Two or more people working together with the idea of generating a profit from a business are known as partners. Although governed by law, partnerships, otherwise known as legally as firms, can be quite informal. The partners agree how the firm should be run, partners share profits and losses in accordance with their agreement.

Like sole trader, a partnership is not legally distinguished from its members. Personal assets of the partners may have to be used to pay the debts of the partnership business.

合夥人的個人財物可能需要用來償付合夥企業的債務。

The advantages of trading as a partnership stem mainly from the there being many owners rather than one. This means that:

· More resources may be available, including capital, specialist knowledge, skills and ideas.

- Administrative expenses may be lower for a partnership than for the equivalent number of sole traders, due to economics of scale;.
- Partners can substitute for each other.

Partners can introduce or withdraw capital at any time, provided that all the partners agree

2.3 Limited Liability Companies

有限責任公司

公司是一個獨立的法人主體。

公司的所有人投入資金並持有公司的股份,享有公司的剩餘權益。

Unlike sole traders and partnerships, companies are established as separate legal entities to their owners. This is achieved through the process of incorporation. The owners of the company invest capital in the business in return for a shareholding that entitles them to a share of the residual assets of the business (i.e., what is left when the business winds up). The shareholders are not personally liable for the debts of the company and whilst they may lose their investment if the company becomes insolvent they will not have to pay the outstanding debts of the company if such a circumstance arises. Likewise, the company is not affected by the insolvency (or death) of individual shareholders.

2.4 Comparison of Companies to Sole Traders and Partnerships

The fact that a company is a separate legal entity means that it is very different from a sole trader or partnership in a number of ways.

(1) Property Holding

有限責任公司的財產歸屬於公司所有。

The property of a limited company belongs to the company. A change in the ownership of shares in the company will have no effect on the ownership of the company's property. (In a partnership the firm's property belongs directly to the partners who can take it with them if they leave the partnership.)

(2) Transferable Shares

公司股份在不經過其他股東的同意下可以依法轉讓。

Shares in a limited company can usually be transferred without the consent of the other shareholders. In the absence of agreement to the contrary, a new partner cannot be introduced into a firm without the consent of all existing partners.

Chapter 1　Introduction to Accounting　5

(3) Suing and Being Sued

As a separate legal person, a limited company can sue and be sued in its own name. Judgments relating to companies do not affect the members personally.

> 有限責任公司可以以自己的名義起訴或被起訴。對公司的判決不影響合夥企業所有人。

(4) Security for Loans

A company has greater scope for raising loans and may secure them with floating charges. A floating charge is a mortgage over the constantly fluctuation assets of a company providing security for the lender of money to a company. It does not prevent the company dealing with the assets in the ordinary course of business. Such a charge is useful when a company has not non-current assets (e.g., land), but does have large and valuable inventories.

> 公司可以通過借款籌資,並提供浮動擔保。
>
> 浮動擔保是指公司抵押其現有的和將來取得的全部資產,為債權人的利益提供擔保。

Generally, the law does not permit partnerships or individuals to secure loans with a floating charge.

(5) Taxation

Because a company is legally separated from its shareholders, it is taxed separately from its shareholders. Partners and sole traders are personally liable for income tax on the profits made by their business.

Unit 3　The Users of Financial Information

3.1　The Need for Financial Statements

There are various groups of people who need information about the activities of a business. Why do businesses need to produce financial statements? If a business is being run efficiently, why should it have to go through all the bother of accounting procedures in order to produce financial information?

The International Accounting Standards Board states in its document *Framework for the preparation and presentation of financial statements* (which we will examine in detail later in this Study Text):

> International Accounting Standards Board 國際會計準則理事會

『The objective of financial statements is to provide information about the financial position, performance and changes in financial position of an entity that is useful to a wide range of users in making economic decisions.』

> 財務報表的目標:提供與企業財務狀況、經營業績和財務狀況變動等相關的信息,有助於報表使用者做出經濟

決策。

In other words, a business should produce information about its activities because there are various groups of people who want or need to know that information. This sounds rather vague, so to make it clearer, we will study the classes of people who need information about a business. We need also to think about what information in particular is of interest to the members of each class.

3.2　Users of Financial Statements and Accounting Information

The following people are likely to be interested in financial information about a large company with listed shares.

企業的管理人員

- **Managers of the company** are appointed by the company's owners to supervise the day-to-day activities of the company. They need information about the company's financial situation as it is currently and as it is expected to be in the future. This is to enable them to manage the business efficiently and to make effective decisions.

股東

- **Shareholders of the company** (i.e., the company' owners) want to assess how well the management is performing. They want to know how profitable the company's operations are and how much profit they can afford to withdraw from the business for their own use.

供應商、顧客

- **Trade contacts** include suppliers who provide goods to the company on credit and customers who purchase the goods or services provided by the company. Suppliers want to know about the company's ability to pay its debts; customers need to know that the company is a secure source of supply and is in no danger of having to close down.

投資人、債權人

- **Investors and creditors** provide the money a business need to get started. When a company opened its first store, the company had no track record. To decide whether to help start a new venture, potential investors evaluate what income they can expect on their investment. This means analysing the financial statements of the business. Before deciding to invest in a company, for example, you may examine the company's financial statements. Before making a loan to the company, banks evaluate the company's ability to meet scheduled payment.

稅務部門	• **The taxation authorities** want to know about business profits in order to assess the tax payable of the company, including sales taxes.
企業職工	• **Employees of the company** should have a right to information about the company's financial situation, because their future careers and the size of their wages and salaries depend on it.
財務分析師、財務顧問	• **Financial analysts and advisers** need information for their clients or audience. For example, stockbrokers need information to advise investors; credit agencies want information to advise potential suppliers of goods to the company; journalists need information for their reading public.
政府機關	• **Government and their agencies** are interested in the allocation of resources and the activities of business entities. They also require information in order to provide a basis for national statistics.
社會公眾	• **The public.** Entities affect members of the public in a variety of ways. For example, they may make a substantial contribution to a local economy by providing employment and using local suppliers.
其他報表使用者	• **Other users.** For example, labour unions demand wages that come from the employer's reported income. Another important factor is the effect of an entity on the environment, for example as regards pollution.

Accounting information is summarised in financial statements to satisfy the information needs of these different groups. Not all will be equally satisfied.

Chapter 2　The Qualitative Characteristics of Financial Information

Accounting practice has developed gradually over time. Many of its procedures are operated automatically by people who have never questioned whether alternative method exist which have equal validity. However, the procedures in common use imply the acceptance of certain concepts which are by no means self-evidence; nor are they the only possible concepts which could be used to build up an accounting framework. In this chapter we shall single out the important assumptions and concepts for discussion.

Unit 1　Fundamental Accounting Assumptions

The **IASB's Conceptual Framework** sets out one important underlying assumptions for financial statements, the **going concern concept**.

持續經營

1.1　Going Concern

The financial statements are normally prepared on the assumption that an entity will continue in operation for the foreseeable future. In other words, the entity has neither the intention or the need to liquidate or curtail materially the scale of its operations.

持續經營最重要的意義就是資產不能使用清算價值計價。

The main significance of the going concern concept is that the assets should not be values at their 『**break-up**』 value.

For example, a retailer commences business on 1 January and buys inventory of 200 T-shirts, each costing ＄10. During the year he sells 140 T-shirts at ＄20 each.

Chapter 2 The Qualitative Characteristics of Financial Information   9

· If the business is regarded as a going concern, the inventory unsold at 31 December will be carried forward into the following year, when the cost of the remaining T-shirts will be matched against the eventual sale proceeds in computing that year's profits. The sixty T-shirts will therefore be valued at $600 (60 × $10).

· If the business is to be closed down, the remaining sixty T-shirts must be valued at the amount they will realize in a forced sale, which is $360 (60 × $6).

1.2 Accrual Basis

權責發生制

Entities should prepare their financial statements on the basis that transactions are recorded in them, not as the cash is paid or received, but as the revenues or expenses are earned or incurred in the accounting period to which they relate.

權責發生制不是會計基本假設。

The accrual basis is not an underlying assumption. However, the Conceptual Framework makes clear that financial statements should be prepared on an accrual basis.

權責發生制下，發生的費用與獲得的收入相配比，並計算利潤。這就是配比原則。

According to the accrual concept, computing profit revenue earned must match expenditure incurred in earning it. This is also known as the **matching convention**.

For example, Emma purchases 20 T-shirts in her first month of trading at a cost of $5 each. She then sells all of them for $10 each. Therefore, Emma has made a profit of $100, by matching the revenue ($200) earned against the cost ($100) of acquiring them.

If Emma only sells 18 T-shirts, it is incorrect to charge her statement of profit or loss with the cost of 20 T-shirts, as she still has two T-shirts in inventory. Therefore, only the purchase cost of 18 T-shirts ($90) should be matched her sales revenue ($180), leaving her with a profit of $90.

Unit 2 The Qualitative Characteristics of Financial Information

Qualitative characteristics are the attributes that make information provided in financial statements useful to others.

The Framework splits qualitative characteristics into two categories:

(a) Fundamental qualitative characteristics.

相關性	· relevance;
真實性	· faithful representation.
	(b) Enhancing qualitative characteristics.
可比性	· comparability;
可核性	· verifiability;
及時性	· timeliness;
可理解性	· understandability.

2.1　Fundamental qualitative characteristics

2.1.1　Relevance

相關性是指信息具有影響使用者決策的能力。

Only relevant information can be useful. Information is relevance if it has the ability to influence the economic decisions of users, and is provided in time to influence those decisions. Information that is relevant has predictive, or confirmatory value.

相關信息具有預測價值和反饋價值。

· Predictive value enables users to evaluate or assess past, present or future events.

· Confirmatory value helps users to confirm or corrects past evaluations and assessments.

信息的相關性也受其特性和重要性的影響。

The relevance of information is affected by its nature and **materiality**.

如果財務報表上的信息遺漏或錯誤地列報影響報表使用者做出經濟決策，則該信息是重要信息。

Information is material if its omission or misstatement could influence the economic decisions of users taken on the basis of the financial statements.

In assessing whether or not an item is material, it is not only the value of the item which needs to be considered. The context is also important.

Example 2.1

If a statement of financial position shows non-current assets of $ 2 million and inventories of $ 30,000, an error of $ 20,000 in the depreciation calculations might not be regarded as material. However, an error of $ 20,000 in the inventory valuation would be material. In other words, the total of which the error forms part must be considered.

Example 2.2

If a business has a $ 85,000 balance on bank deposit account and a bank loan of $ 80,000, it will be a material misstatement if these two amounts are displayed on the statement of financial position as 〖cash at bank $ 5,000〗. In other words, incorrect presentation may amount to material misstatement even if there is no monetary error.

2.1.2 Faithful Representation

If information is to represent faithfully the transactions and other events that it purports to represent, they must be accounted for and presented in accordance with their substance and economic reality, and not merely their legal form. This is known as 〖substance over form〗.

信息應當按照經濟實質進行列報，而不應當僅僅按照其法律形式。這就是「實質重於形式」。

To be a perfectly faithful representation, information must be **complete**, **neutral** and **free from error**.

(1) Completeness

To be understandable information must contain all the necessary descriptions and explanations.

可理解的信息應當包含所有的必要說明和解釋。

(2) Neutrality

Information must be neutral (i.e., free from bias) in the selection or presentation of financial information. This means that information must not be manipulated in any way in order to influence the decisions of users.

信息應當具有中立性。

(3) Free from Errors

正確無誤

Information must be free from error within the bounds of materiality. A material error or an omission can cause the financial statements to be false or misleading and thus unreliable and deficient in terms of their relevance. Free from error does not mean perfectly accurate in all respects. For example, where an estimate has been used the amount must be described clearly and accurately as being estimate.

2.2 Enhancing Qualitative Characteristics

Comparability, verifiability, timeliness and understandability are qualitative characteristics that enhance the usefulness of information which is relevant and faithfully represented.

2.2.1 Comparability

In order to make decisions, users need to compare information between entities and over a time period. The information from different entities is comparable if there is consistency in the accounting treatment of the economic events and transactions over time and in the disclosure of accounting policies.

An important implication of comparability is that users are informed of the accounting policies employed in preparation of the financial statements, any changes in those policies and the effects of such changes. Compliance with accounting standards, including the disclosure of the accounting policies used by the entity, helps to achieve comparability.

2.2.2 Verifiability

可核性是指信息可以直接驗證，如現金盤點；也可以間接核驗，通過模型、公式和其他技術手段加以計算。

Verifiability can be direct or indirect. Direct verifiability means verifying an amount or other representation through direct observation (i.e., counting cash). Indirect verification means checking the inputs to a model, formula or other technique and recalculation the outputs using the same methodology.

2.2.3 Timeliness

Timeliness means having information available to decision makers in time to be capable of influencing their decisions. Generally, the older information becomes less useful.

經濟業務發展過程中，及時披露信息，可能導致信息不完整或出現偏差；反之，如果披露不及時，信息可能變得不相關。信息質量特性衝突矛盾時，優先考慮能夠滿足報表使用者做出經濟決策需求的質量特性。

If information is reported on a timely basis when not all aspects of the transaction are known, it may not be complete or free from error. Conversely, if every detail of a transaction is known, it may be too late to publish the information because it has become irrelevant. The overriding consideration is how best to satisfy the economic decision-making needs of the users.

2.2.4 Understandability

可理解性取決於：
· 信息的列報方式。
· 使用者理解信息的能力。

Understandability depends on:
· The way in which information is presented.
· The capabilities of users.

It is assumed that users:

2.3 Other Accounting Concepts

There are other accounting concepts which are useful in the preparation of financial statements.

2.3.1 Fair Presentation

公允的列示是指在編製財務報表時，應遵循適用的財務報告準則及相關的法律法規。

Fair presentation relates to preparation of financial statements in accordance with applicable financial reporting standards, relevant laws and regulations.

Disclosure of compliance with reporting standards should be disclosed in the financial statements. If there is less than full compliance, the extent of non-compliance should be disclosed and explained.

2.3.2 Consistency

企業不同時期發生的相似的交易或者事項，應採用一致的會計處理方法和政策。

It is assumed that the enterprise will apply the same accounting treatment and policies for similar transactions and events, and these policies and principles will be adopted from year to year.

For example, if the entity depreciates a building using the straight-line method; then all similar buildings will also be depreciated using the straight-line method. The same depreciation method is used from one accounting year to another. Application of consistency allows financial statements to be comparable.

2.3.3 The Business Entity Concept

會計主體原則

Financial statements always treat the business as a separate entity.

財務報表中的財務信息只能反應企業本身的生產經營活動，與所有者個人的經濟活動無關。

This principle means that the financial accounting information presented in the financial statements relates only to the activities of the business and not to those of the owner. From an accounting perspective the business is treated as being separate from its owners.

Chapter 3 Source Document, Records and Books of Prime Entry

Accounting is the process of analyzing, recording, classifying, summarizing, reporting, and interpreting information. Financial data enters the accounting process in the form of transactions (financial events). The flow of information from the initial transaction to the financial statements is illustrated as follows:

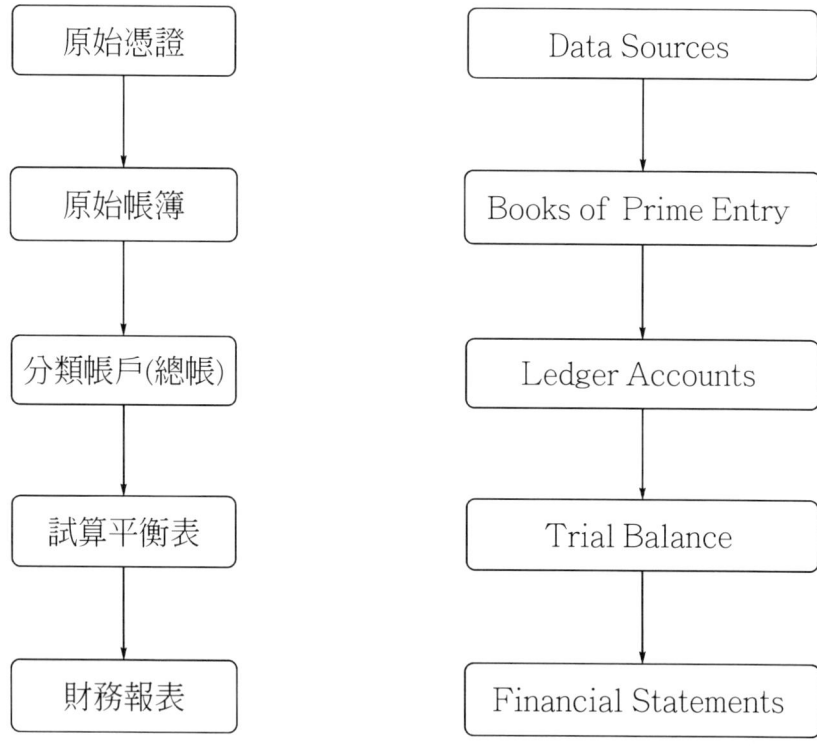

Chapter 3 Source Document, Records and Books of Prime Entry   15

Unit 1 Types of Source Documents

In every business a number of transactions and events will take place every day. The main transactions that take place include sales and purchases (goods and services). Others include rental costs, raising finance, repayment of finance, etc. All of these transactions must be adequately captured by the financial reporting system.

Whenever a business transactions takes place, involving sales or purchases, receiving or paying money, owing or being owed money, it is usual for the transaction to be recorded on a document. These documents are the source of all the information recorded by a business.

報價單　　(1) Quotation

A document sent to a customer by a company stating the fixed price that would be charged to produce or deliver goods and services. Quotations tend to be used when businesses do not have a standard listing of prices for products, for example when the time, materials and skills required for each job vary according to the customer's needs. Quotations can't be changed once they have been accepted by the customer.

購買訂單　　(2) Purchase Order

A document of the company that details goods or services which the company wishes to purchase from another company. Two copies of a purchase order are often made. One is sent to the company from which the goods or service will be purchased, and the other is kept internally so the company can keep track of its order.

銷貨訂單　　(3) Sales Order

A document of the company that details an order placed by a customer for goods or services. The customer may have sent a purchase order to the company from which the company will then generate a sales order.

Sales orders are usually sequentially numbered so that the company can keep track of orders placed by customers

發貨通知單　　(4) Goods Dispatched Note (GDN)

A document of the company which lists the goods that the company has sent out to a customer. The company will keep a record of goods dispatched notes in case of any queries by customers about goods sent. The customer will compare the goods dispatched note so that they received to make sure all the items listed have been delivered and are the right specification.

收貨單　　（5）Goods Received Note (GRN)

A document of the company that record of goods received at the point of receipt. This record is used to confirm all goods have been received and often compared to a purchase order before payment is issued.

發票　　（6）Invoice

A document issued by supplier of goods as a request for payment. For the supplier, selling the goods or services will be treated as a sales invoice. For the customer, this will be treated as a purchase invoice.

對帳單　　（7）Statement

A document sent out by a supplier to a customer listing the transactions on the customer's account, including all invoices and credit notes issued, and all payments received from the customer. The statement is useful as it ensure that the amount owing is correct. Any differences can be queried.

貸項通知單　　（8）Credit Note

A document sent by a supplier to a customer in respect of goods returned or overpayments made by the customer. It is a 『negative』invoice.

借項通知單　　（9）Debit Note

A document sent by a customer to a supplier in respect of goods returned or an overpayment made.

It is a formal request for the supplier to issue a credit note.

匯款通知單　　（10）Remittance Advice

A document sent to a supplier or as notification of a payment, detailing which invoice are being paid and which credit notes offset.

現金收據　　（11）Receipt

A document issued by the selling company that indicate the payment has been received. This is usually in respect of cash sales (e.g., a till receipt from a cash register).

Unit 2 Books of Prime Entry

In the course of business, source documents are created. The details on these source documents need to be summarized, otherwise the business might forget to ask for some money, or forget to pay some, or even accidentally pay something twice. It needs to keep records of source documents of transactions, so that it known as what is going on. Such records are made in books of prime entry.

原始憑證被分類記入不同的原始帳簿。

Several books of prime entry exist, each recording a different type of transaction. Look at Table 3.1.

Table 3.1

Books of prime entry	原始帳簿
Sales day book	銷售日記帳
Purchases day book	採購日記帳
Sales returns day book	銷售退回日記帳
Purchases returns day book	採購退回日記帳
Cash book	銀行存款日記帳
Petty cash book	現金日記帳
Journal	普通日記帳
Transaction type	**經濟業務的類型**
Credit sales	賒銷業務
Credit purchases	賒購業務
Returns of goods sold on credit	賒銷退回
Returns of goods bought on credit	賒購退回
All bank transactions	銀行存款收付業務
All small cash tractions	零星現金收付業務
All transactions not recorded elsewhere	不記入上述帳簿中的其他經濟業務

2.1 Sales Day Book

銷售日記帳是記錄賒銷業務的原始帳簿。

現金銷售業務記錄在銀行存款日記帳。

The sales day book is the book of prime entry for credit sales (i.e., the goods are sold and payment is collected at a later date). The sales day book is used to keep a list of all invoices sent out to customers each day. Cash sales are recorded in the cash book.

Table 3.2 shows an extract from a sales day book might look like this.

Table 3.2   **Sales Day Book**

Date	Invoice	Customer	Total amount invoiced $
06/09/2016	247	Jake	105.00
	248	Bella	86.40
	249	Milo	31.80
	250	Max	1,264.60
			1,487.80

The total sales for the day of $ 1487.80 will be entered into the accounting ledgers in double entry format.

2.2 Purchase Day Book

採購日記帳匯總記錄企業發生的賒購業務。現金採購業務記錄在銀行存款日記帳。

The purchase day book summarises the daily purchases made on credit terms. Cash purchases are recorded in the cash book.

Table 3.3 shows an extract from a purchase day book might look like this.

Table 3.3   **Purchases Day Book**

Date	Supplier	Total amount invoiced $
10/09/2016	Harry	315.30
	Ron	29.40
	Draco	116.80
	Neville	100.50
		562.00

Chapter 3　Source Document, Records and Books of Prime Entry　19

The total purchases for the day of $562.00 will be entered into the accounting ledgers in double entry format.

Note that:

There is no 「invoice number」 column, because the purchase day book records other people's invoice, which have all sorts of different numbers.

2.3　Sales Returns Day Book

所有的貸項通知單記錄在銷售退回日記帳中。

When customers return goods for some reason, a credit note is raised. All credit notes are recorded in the sales returns day book.

Table 3.4 shows an extract from the sales returns day book.

Table 3.4　　　　　　　　　　**Sales Returns Day Book**

Date	Credit note	Customer	Amount
			$
30/04/2016	CR23	Sam	164.50

A business with very few sales returns may record a credit note as a negative entry in the sales day book.

2.4　Purchase Returns Day Book

採購退回日記帳記錄由於商品退回給供應商而收到的貸項通知單。

The purchase returns day book records credit notes received in respect of goods which the business sends back to its suppliers.

Table 3.5 shows an extract from the purchase returns day book.

Table 3.5　　　　　　　　　**Purchase Returns Day Book**

Date	Supplier	Amount
		$
30/04/2016	Harry	325.60

Once again, a business with very few purchase returns may record a credit note received as a negative entry in the purchase day book.

2.5 Cash Book

All transactions involving cash at bank are recorded in the cash book. Many businesses have two distinct cash books — a cash payments book and a cash receipts book. It is common for businesses to use a columnar format cash book in order to analyse types of cash payment and receipt.

很多企業有兩本不同的銀行存款日記帳——銀行付款日記帳和銀行收款日記帳。

The receipts part of the cash book would look like Table 3.6 shown below.

Table 3.6 **Cash Book (Receipts)**

Date	Narrative	Total	Sales	Receivables
01/06/2016	Balance b/d	900		
	Accounts receivable: C Monet	1,500.00		1,500.00
	Accounts receivable: W Gogh	3,000.00		3,000.00
	Cash sale	150.00	150.00	
	Sale of non-current asset	600.00		
		5,550.00	150.00	4,500.00

Table 3.7 **Cash Book (Payments)**

Date	Narrative	Total	Purchase	Payables	Rent
01/06/2016	Cash purchase	1,400.00	1,400.00		
	Accounts payable: Mr A	1,600.00		1,600.00	
	Accounts payable: Mr B	800.00		800.00	
	Rental	1,000.00			1,000.00
	Balance c/d	750.00			
		5,550.00	1,400.00	2,400.00	1,000.00

2.6 Petty Cash Book

現金日記帳是記錄零星現金收支的日記帳。

A petty cash book is a cash book for small payments. Although the amounts involved are small, petty cash transactions still need to be recorded. Otherwise, the cash float could be abused for personal expenses or even stolen.

The cash receipts will be recorded together with the payments which will be analysed in the same way as a cash book. There are usually more payments than receipts, and petty cash must be 「topped up」 from time to time with cash from the business bank account. A typical layout looks as Table 3.8 shows.

Table 3.8   **Petty Cash Book**

Receipt	Date	Narrative	Total	Postage	Milk	Travel
$ 250	1/09/20×6	Balance b/d	$	$	$	$
		Taxi fare	25			25
		Postage stamps	10	10		
		Milk bill	28		28	
		Balance c/d	187			
250			250	10	28	25

在定額備用金制度下，現金保持定額數，因此現金的補足金額與現金的使用數額相等。

An **imprest system** will be adopted for the petty cash book. Under the imprest system, the petty cash is kept at an agree sum, so that each topping up is equal to the amount paid out in the period.

2.7 The Journal

The journal is the record of prime entry for transactions which are not recorded in any of the other books of prime entry.

The journal is used to record:

·差錯更正；

·沖銷壞帳；

·計提折舊；

·不能記入其他日記帳中的其他事項。

· the correction of errors;

· writing off bad or irrecoverable debts;

· depreciation charge;

· other items not recorded in other books of prime entry.

Note that:

The presentation and the narrative explanation of the transaction, which is a requirement for all journal entries.

Table 3.9 is an example of journal.

Table 3.9   **Journal**

31 January	Debit	Depreciation expense	$ 3,650	
	Credit	Accumulated depreciation		$ 3,650
	With the depreciation charge for the period			
31 January	Debit	Purchases	$ 250	
	Credit	Trade accounts payable		$ 250
	A transaction previously omitted			

Chapter 4　Ledger Accounts & Control Account

In most companies each class of transaction and their associated assets and liabilities are given their own account. There will be separate accounts for purchases, sales, inventory assets, amount due from customers, liabilities to pay suppliers, rent, etc. Each account in the system is referred to as a 『ledger』.

簡單地說，帳戶就是運用復式記帳系統記錄所有的經濟業務和事項的方式。

In simple terms, the ledger accounts are where the double entry records of all transactions and events are made. They are the principal books or files for recording and totaling monetary transactions by account. A company's financial statements are generated from summary totals in the ledgers.

Unit 1　General Ledger

1.1　Ledger Accounts

分類帳戶匯總記錄了原始記錄簿中的每一項經濟業務。

Ledger accounts summarise all the individual transactions listed in the books of prime entry. A business should keep a record of the transaction that it makes, the assets it acquires and liabilities it occurs. When the time comes to prepare a statement of profit or loss and a statement of financial position, the relevant information can be taken from those records.

The records of transactions, assets and liabilities should be kept in the following ways:

　· In chronological order and dated, so that transactions can be related to a particular period of time.

　· Built up in cumulative totals.

帳戶的結構

1.2　The Format of a Ledger Account

每一個帳戶都有兩個方向和帳戶名稱，稱為T形帳戶。
　· 帳戶上方是帳戶的名稱。
　· 左方，又稱借方。

There are two sides to a ledger account and an account heading on top, so they are often referred to as T accounts.
　· On top of the account is its name.
　· There is a left hand side, or debit side.

· 右方，又稱貸方。　　　　　· There is a right hand side, or credit side.

Name of account

Debit side	$	Credit side	$

1.3 The Nominal Ledger

The nominal ledger or general ledger is an accounting record which summarises the financial affairs of a business. It contains details of assets, liabilities, capital, income and expenditure, profit and loss. It consists of a large number of different accounts, each account having its own purpose or 『name』and an identity or code. Examples of accounts in the nominal ledger include the following:

· 廠房及設備原值　　　　　· plant and machinery at cost (non-current asset);

· 機動車輛原值　　　　　　· motor vehicles at cost (non-current asset);

· 廠房及設備，計提折舊　　· plant and machinery, provision for depreciation (non-current asset);

· 機動車輛，計提折舊　　　· motor vehicles, provision for depreciation (non-current asset);

· 所有者資本　　　　　　　· proprietor's capital (equity);

· 存貨——原材料　　　　　· inventory – raw material (current assets);

· 存貨——產成品　　　　　· inventory – finished goods (current assets);

· 應收帳款　　　　　　　　· total trade accounts receivable (current assets);

· 應付帳款　　　　　　　　· total trade accounts payable (current liability);

· 工資與薪金　　　　　　　· wages and salaries (expense item);

· 租金　　　　　　　　　　· rent (expense item);

· 廣告費用　　　　　　　　· advertising expenses (expense item);

· 銀行手續費　　　　　　　· bank charges (expense item);

· 電話費　　　　　　　　　· telephone expenses (expense item);

· 銷售收入　　　　　　　　· sales (revenue item);

· 現金或銀行透支　　　　　· total cash or bank overdraft (current assets or liability).

收入和費用帳戶形成利潤表，資產和負債帳戶形成資產負債表。

When it comes to drawing up the financial statements, the revenue and expense accounts will help to form the statement of profit or loss, while the asset and liability accounts go into the statement of financial position.

Unit 2　Control Accounts

控制帳戶是總分類帳簿中用於記錄同類經濟業務總金額的帳戶。控制帳戶主要用於應收帳款和應付帳款。

A control account is an account in the nominal ledger in which a record is kept of the total value of a number of similar but individual items. Control accounts are used chiefly for trade receivables and payables.

The reasons for having control accounts are as follows:

·檢驗應收帳款明細帳和應付帳款明細帳記錄是否正確。

· They provide a check on the accuracy of entries made in the personal accounts in the receivables ledger and payables ledger.

·便於查找差錯。

· The control accounts also assist in the location of errors, where posting to the control accounts are made daily or weekly, or even monthly.

·便於分清職能，有利於內部查驗。

· Where there is a separation of clerical duties, the control account provides an internal check.

·編製試算平衡表或財務報表時，可以更快地提供應收及應付帳款的餘額。

· To provide total receivables and payables balances more quickly for producing a trial balance or statement of financial position.

2.1　Receivables Control Account

A receivable control account is an account in which records are kept of transactions involving all receivables in total. The balance on the receivables control account at any time be the total amount due to the business at that time from its receivables.

2.1.1　Preparing the Control Account

Transactions affecting credit customers are:

·銷售發票；
·貸項通知單；
·客戶支付的款項；
·客戶享有的現金折扣。

· sales invoices;
· credit notes for sales returns;
· payments by customers;
· settlement discounts taken by customers (will be described in a later chapter).

每隔一段時間，一個星期或一個月，根據銷售日記帳記錄的賒銷總額編製分錄。

借：應收帳款控制帳戶
貸：銷售收入

At regular intervals, perhaps weekly or monthly, the total for sales on credit according to the sales day book is recorded.

Dr　Receivables control account
Cr　Sales

之後，銷售日記帳中的每一項銷售業務將被記入相應的明細分類（個人）帳戶。

Immediately afterwards, the individual sales listed in the sales day book are charged to the appropriate personal accounts.

Example 4.1

Greg Tyson is a supplier of widgets. He has three regular credit customers. At the beginning of July, none of these customers owed any money. Transactions with these customers during July were recorded in the books of prime entry as Table 4.1 shows.

Table 4.1   **Sales Day Book**

Date	Invoice No.	Customer	Total amount invoiced
July			$
1	3031	Able	846.00
6	3032	Baker	940.00
8	3033	Charlie	258.50
			2,044.50

Table 4.2   **Cash Receipts Book (extract)**

Date	Detail	Receivables
July		$
20	Able	500.00
25	Baker	390.00
		890.00

上述經濟業務的總額被記入應收帳款控制帳戶。

The totals of these transactions would be posted to the receivables control account as follows:

Receivables Control Account

	$		$
Sales	2,044.50	Bank	890.00

每一筆交易被記入明細帳戶（個人帳戶）。

In additional to posting the totals to the receivables control account, the individual transaction details are posted to the personal accounts as follows:

Able Account

	$		$
Sales Inv 3031	846.00	Bank	500.00

Baker Account

	$		$
Sales Inv 3032	940.00	Bank	390.00

Charlie Account

	$		$
Sales Inv 3033	258.5		

The ledger accounts can then be balanced and reconciled. The reconciliation of the control account total balance to the total balances of the individual customer accounts is shown below.

2.1.2 Receivables Control Account Reconciliation

For each account, the closing balance is the difference between the total debits and the total credits.

Receivables Control Account

	$		$
Sales	2,044.50	Bank	890.00
		Balance c/d	1,154.50
	2,044.50		2,044.50
Balance b/d	1,154.40		

餘額結轉下期；上期結轉餘額。

Note: c/d: carried forward; b/d: brought forward.

Able Account

	$		$
Sales Inv 3031	846.00	Bank	500.00
		Balance c/d	346.00
	846.00		846.00
Balance b/d	346.00		

Baker Account

	$		$
Sales Inv 3032	940.00	Bank	390.00
		Balance c/d	550.00
	940.00		940.00
Balance b/d	550.00		

Charlie Account

	$		$
Sales Inv 3033	258.50	Balance c/d	258.50
	258.50		258.50
Balance b/d	258.50		

Table 4.3   **Reconciliation Statement**

	$
Individual customers' balances	
Able	346.00
Baker	550.00
Charlie	258.50
Total of individual account balances	1,154.50
Balance on control account	1,154.50
discrepancy	0

應收帳款個人（明細）帳戶的餘額總和應與其控制帳戶餘額相符。

Here, the total of the individual receivables account balances agrees with the balance on the control account. This should be expected, so there appears to be no error.

2.2 Payables Control Account

A payables control account is an account in which records are kept of transactions involving all payables in total. The balance on this account at any time will be the total amount owed by the business at that time to its payables.

2.2.1 Preparing the Control Account

Transactions with suppliers are recorded in both the books of prime entry and the payables ledger. The transactions relating to credit purchases from suppliers are:

- 採購發票;
- 採購退回收到的貸項通知單;
- 支付給供應商的款項;
- 供應商給予的現金折扣。

- purchase invoice;
- credit notes for purchase returns;
- payments to suppliers;
- settlement discounts taken from suppliers (will be described in later chapter).

原始記錄簿中的信息被過入應付帳款分類帳戶:
借: 採購成本
貸: 應付帳款控制帳戶
此外, 每一項採購業務將被過入相應的明細分類(個人)帳戶。

This information are entered in book of prime entry, and then posted to the payables ledgers as follows:

Dr Purchase
Cr Payables control account

In addition to posting the totals to the payables ledger, the value of individual transactions is posted to the personal accounts.

Example 4.2

Bob James is a car parts distributor. He has four credit suppliers. At the beginning of July, he did not owe any of these suppliers money. Transactions with these suppliers during July were recorded in the books of prime entry as Table 4.4 shows.

Table 4.4   **Purchases Day Book**

Date	Supplier	Supplier account number	Total
July			$
1	Ace	6031	1,692.00
1	Bays	6032	1,880.00
8	Eastern	6033	1,480.50
10	Dans	6034	968.00
			6,020.50

Table 4.5 **Cash Payments Book (extract)**

Date	Detail	Payables
July		$
20	Ace	1,692.00
20	Bays	1,000.00
25	Eastern	1,500.00
		4,192.00

經濟業務的總金額將被過入應付帳款控制帳戶。

The total of these transactions would be posted to the payables control account as follows:

Payables Control Account

	$		$
Bank	4,192.00	Purchases	6,020.50

每一項經濟業務信息被過入個人帳戶（明細分類帳戶）。

In addition to posting the totals to the control account, the individual transaction details are posted to the personal accounts as follows:

Ace Account

	$		$
Bank	1,692.00	Purchases	1,692.00

Bays Account

	$		$
Bank	1,000.00	Purchases	1,880.00

Eastern Account

	$		$
Bank	1,500.00	Purchases	1,480.50

2.2.2 Payables Control Account Reconciliation

Reconciliation is very similar to the approach with the receivables control account. At any time, the balance on the control account should be equal to the total of all the balances on the individual supplier accounts in the payables ledger.

The reconciliation of payables control account is shown below.

Payables Control Account

	$		$
Bank	4,192.00	Purchases	6,020.50
Balance c/d	1,828.50		
	6,020.50		6,020.50
		Balance b/d	1,828.50

Ace Account

	$		$
Bank	1,692.00	Purchases	1,692.00
	1,692.00		1,692.00

Bays Account

	$		$
Bank	1,000.00	Purchases	1,880.00
Balance c/d	880.00		
	1,880.00		1,880.00
		Balance b/d	880.00

Eastern Account

	$		$
Bank	1,500.00	Purchases	1,480.50
		Balance c/d	19.50
	1,500.00		1,500.00
Balance b/d	19.5		

注意，Eastern 的帳戶餘額在借方，這是因為 Bob James 支付的金額超過他的欠款。

Note that there is a debit balance on Eastern's account. This is because Bob James has paid more than the amount of his debt, and Eastern therefore owes money to Bob James. Eatern's account is 『in debit』.

Dans Account

		$			$
	Balance c/d	968.00		Purchases	968.00
		968.00			968.00
				Balance b/d	968.00

Table 4.6   **Reconciliation Statement**

	$
Individual suppliers' balances	
Ace	0
Bays	880.00
Eastern (minus value, because debit balance)	(19.50)
Dans	968
Total of individual account balances	1,828.50
Balance on control account	1,828.50
Discrepancy	0

應付帳款個人（明細）帳戶的餘額總和與其總帳帳戶相符，則沒有記帳差錯。

Here, the total of the individual payables account balances agrees with the balance on the control account. This should be expected, so there appears to be no error.

Chapter 5　Double Entry Bookkeeping

復式記帳法是將原始記錄簿中的會計信息過入到總分類帳戶時使用的一種方法。

Double entry bookkeeping is the method used to transfer accounting information from the books of prime entry into the nominal ledger. It is the cornerstone of accounts preparation and is surprisingly simple.

每項經濟業務都會從兩個方面影響企業的財務報表。

Each transaction that a business enters into affects the financial statements in two ways. To follow the rules of double entry bookkeeping, each time a transaction is recorded, both effects must be taken into account. These two effects are equal and opposite, as such, the accounting equation will always be maintained.

Unit 1　Accounting Elements and Accounting Equation

1.1　Accounting Elements

Before the accounting process can begin, the organization must be defined. A business entity could be an individual, association, or other organization that engages in business activities. This definition separates personal from business finances of an owner. A business that is owned by one person is called a proprietorship. The owner of the business is known as the proprietor. Here, we will only discuss proprietorship.

財務報表要素：資產、負債、所有者權益、費用和收入。

In order to appropriately report the financial performance and position of a business, the financial statements must summarise five key elements: assets, liabilities, owner's equity, expenses and income.

1.1.1　Assets

資產是指過去的交易形成的，由企業控制的，可望未來經濟利益流入企業的經濟資源。

An asset is a resource controlled by the entity as a result of past events from which future economic benefits are expected to flow to the entity. Examples of assets are as follows:

- land and buildings owned by the business;
- plant and machinery;
- office equipment;
- motor vehicles;
- stocks of materials or goods for resale;
- money in a business bank account;
- notes and coins (including 『petty cash』).

1.1.2 Liabilities

負債是指過去的交易或事項形成的，導致經濟利益流出的現時義務。

A liability is an obligation to transfer economic benefit as a result of past transactions or events. Examples of liabilities are as follows:

- money owed to suppliers for goods and services;
- bank loan;
- bank overdraft;
- money owed to the tax authorities.

1.1.3 Owner's Equity

所有者權益是指企業資產扣除負債後的剩餘權益。

This is the 『residual interest』 in a business and represents what is left when the business is wound up, all the assets sold and all the outstanding liabilities paid. It is effectively what is paid back to the owners (shareholders) when the business ceases to trade. Four main types of transactions affect the owner's interest in the entity:

- investment of assets in the entity by owner;
- withdrawal of assets by the owner;
- income derived;
- expenses incurred.

1.1.4 Income

收入就是企業在報告期內確認的經濟利益流入。

This is the recognition of the inflow of economic benefit to the entity in the reporting period. This can be achieved, for example, by earning sales revenue or though the increase in value of an asset.

收入按性質分類，分為銷售商品收入、提供勞務收入、諮詢收入、佣金收入等。

Revenue represents income which arises in the course of the ordinary activities of an entity. Revenues are classified into many categories depending on their nature (e.g., sales revenue, services revenue, consulting fees revenue, commission revenue).

1.1.5 Expenses

This is the recognition of the outflow of economic benefit from an entity in the reporting period. Examples of expenses include:

- the cost of salaries and wages for employees;
- telephone charges and postage costs;
- the rental cost on a building used by the business;
- the interest cost on a bank loan.

1.2 Accounting Equation

The accounting system reflects two basic aspects of a business enterprise: what it owns and what it owes. Liabilities and equity are the source of funds to acquire assets. The financial condition or position of a business is represented by the relationship of assets, liabilities and owner's equity, and is reflected in the following accounting equation.

Assets = Liabilities + Owner's Equity

This equation shows assets are equal to equities. Equities are divided into liabilities and owner's equity. The reason this equality is maintained is that the assets controlled by the business had to be financed by someone, either the owner or lenders. Conversely, the input from the owner and lender has to be represented by asset or assets to the same value. Thus the accounting equation expresses simply that at any point in time the assets of the business will be equal to its liabilities plus the equity of the business.

Example 5.1

The transactions of a new business in its first five days are as follows:

Day 1: Avon commences business introducing $1,000 cash.

Day 2: Buys a motor car for $400 cash.

Day 3: Obtains a $1,000 loan.

Day 4: Purchases goods for $300 cash.

Day 5: Sells goods for $400 on credit.

Use the accounting equation to illustrate the position of the business at the end of each day (ignore inventory for this example).

Solution

Day 1: Avon commences business introducing $1,000 cash

The dual effect of this transaction is as follows:

- The business has $1,000 of cash.
- The business owes the owner $1,000 – this is owner's equity.

Assets = Liabilities + Owner's Equity

$	$	$
1,000	0	1,000

Day 2: Buys a motor car for $400 cash.

The dual effect of this transaction is as follows:

- The business has an asset of $400.
- The business has spent $400 in cash.

Assets = Liabilities + Owner's Equity

$	$	$
1,000	0	1,000
(400)	0	0
400		
1,000	0	1,000

Day 3: Obtains a $1,000 loan.

The dual effect of this transaction is as follows:

- The business has $1,000 of cash.
- The business owes $1,000 to the bank.

Assets = Liabilities + Owner's Equity

1,000	0	1,000
1,000	1,000	0
2,000	1,000	1,000

Day 4: Purchases goods for $300 cash.

The dual effect of this transaction is as follows:

- The business has an expense of $300 (expenses reduce the amount due to the owners (i.e., they reduce equity).
- The business has reduced cash by $300.

Assets＝Liabilities＋Owner's Equity

$	$	$
2,000	1,000	1,000
(300)	0	(300)
1,700	1,000	700

Day 5: Sells goods for $400 on credit.

The dual effect of this transaction is as follows:

· The business has earned sales revenue of $400.

· The business has a new asset to receive payment of $400 from their customer.

Assets＝Liabilities＋Owner's Equity

$	$	$
1,700	1,000	700
400	0	400
2,100	1,000	1,100

Unit 2 Introduction of Double Entry Bookkeeping

復式記帳是企業記錄經濟業務時使用的一種方法。

每一筆會計業務都必須同時記入某一個分類帳戶的借方，並以相等的金額記入對應帳戶的貸方。

Double entry bookkeeping is the method by which a business records financial transactions. It is based on the idea that each transaction has an equal but opposite effect. Every accounting event must be entered in ledger accounts both as a debit and as an equal but opposite credit.

2.1 Debits and Credits

每一個帳戶都有兩個方向，借方和貸方。借方在帳戶的左邊，貸方在帳戶的右邊。

Each account has two sides, a debit side and a credit side. By convention, the debit side is shown on the left and the credit side on the right. For practical purposes, it is both clearer and more convenient to think of the words debit and credit as meaning the left-hand and right-hand sides of the page respectively.

帳戶名稱	
借方　　$	貸方　　$

Name of Account	
Debit side　　$	Credit side　　$

You might see that the account looks a bit like the letter 『T』, which is why you might hear manual accounts drawn up in this way referred to as 『T-accounts』.

A recognised shorthand form is used for debit and credit.

Dr = Debit

Cr = Credit

2.2　The Basic Rules of Double Entry Bookkeeping

每一項經濟業務都會形成兩個會計記錄，一個是借方記錄，另一個是貸方記錄。

The basic rule, which must always be observed, is that every financial transaction gives rise to two accounting entries, one is a debit and the other is a credit. There are five types of account:

- asset;
- liability;
- owner's equity;
- income;
- expense.

Every transaction will affect two accounts because of the dual aspect.

資產、費用增加記在帳戶的借方。
負債、所有者權益和收入增加記在帳戶的貸方。

By convention, an increase in an asset or an expense is recorded on the debit (or left-hand) side of that item's account. Assets and expenses are the opposites of liabilities, owner's equity and income, so an increase in a liability, owner's equity or income balance is recorded on the credit (or right-hand) side. It follows that decreases in assets or expenses are recorded on the credit side, while decreases in liabilities, owner's equity or income are recorded on the debit side. If more than one account is used to record a transaction, the total value of the debit entries and the total value of the credit entries for the transaction must be equal.

The following rules of double entry accounting apply:

借方記錄：
- 資產增加；
- 負債減少；
- 費用增加。

貸方記錄：
- 資產減少；

A debit entry will:
- increase an asset;
- decrease a liability;
- increase an expense.

A credit entry will:
- decrease an asset;

- 負債增加；
- 收入增加。

- increase a liability;
- increase income.

In terms of 『T』 accounts:

資產	
借方	貸方
增加	減少

Assets	
Debit side	Credit side
Increase	Decrease

費用	
借方	貸方
增加	減少

Expenses	
Debit side	Credit side
Increase	Decrease

負債	
借方	貸方
減少	增加

Liabilities	
Debit side	Credit side
Decrease	Increase

所有者權益	
借方	貸方
減少	增加

Owner's Equity	
Debit side	Credit side
Decrease	Increase

收入	
借方	貸方
減少	增加

Income	
Debit side	Credit side
Decrease	Increase

Unit 3 The Use of Double Entry Bookkeeping

The basic rules of double entry bookkeeping can be illustrated using the following transactions. The date of each transactions is not shown here, but in practice transaction dates would be recorded in the accounts.

3.1 Use of Asset, Liability, and Owner's Equity Accounts

Transaction (a): Ted Andy started a business by investing $30,000 in cash.

Cash	
(a)	$ 30,000

Ted Andy, Capital

	(a) $ 30,000

分析：

現金和 Andy 的資本在此經濟業務中增加。現金是一項資產，資產在會計等式的左邊，因此現金帳戶借方增加。Ted Andy 的資本，在等式的右邊，因此 Ted Andy, 資本帳戶貸方增加 $ 30,000。

Analysis

The cash and Andy's captial in the business increased. Since cash is an asset and assets are on the left side of the accounting equation, the cash account was increased by a debit. Andy's capital, is on the right side of the equation and to shows an increase in this account, Ted Andy, capital, was credited for $ 30,000.

Transaction (b): Andy purchased office equipment for $ 2,500 on account.

Office Equipments

(b) $ 2,500	

Accounts Payable

	(b) $ 2,500

分析：

辦公設備，資產類帳戶，借記 $ 2,500。因為負債在等式的右邊，負債增加記入帳戶的右方。企業負債增加，應付帳款帳戶貸記 $ 2,500。

Analysis

An increase in the asset account, office equipment was debit for $ 2,500. Since liabilities are on the right side of the equation, increase to liabilities are shown on the right or credit side of the account. To increase the liabilities of the business, accounts payable was credited for $ 2,500.

Transaction (c): Andy purchased office supplies for cash, $ 350.

Cash

(a) $ 30,000	(c) $ 350

Office Supplies

(c) $ 350	

分析：

Analysis

當一項資產增加，則另外一項資產減少，因此資產總額不變。辦公用具增加 $ 350 記借方，現金減少 $ 350 記貸方。辦公用具是一項資產，雖然使用後將轉化為企業的費用。

One asset was increased while another asset was decreased. There is no change in total asset. Office supplies was debited for the increase and cash was credited for the decrease of $ 350. Office supplies are an asset at the time of purchase even though they will become an expense when used. The procedure used in accounting for supplies used will be discussed later.

Transaction (d): Andy paid $ 500 on account to the company from which the office equipment was purchased.

	Cash		
(a)	$ 30,000	(c)	$ 350
		(d)	$ 500

	Accounts Payable		
(d)	$ 500	(b)	$ 2,500

分析：

應付帳款減少 $ 500，記入借方；現金資產減少 $ 500，記入貸方。

Analysis

The liability accounts payable was decreased with a debit, and the asset cash was decreased with credit for $ 500.

Transaction (e): Purchased office supplies on account, $ 400.

	Office Supplies	
(c)	$ 350	
(e)	$ 400	

	Accounts Payable		
(d)	$ 500	(b)	$ 2,500
		(e)	$ 400

分析：

辦公用具增加，記入借方。應付帳款增加，記入貸方。資產在會計等式的左邊，因此資產的增加記入帳戶的借方。負債在會計等式的右邊，因此負債的增加記入帳戶的貸方。

withdraw（提款）

Analysis

The asset office supplies was increased with a debit. The liability accounts payable was increased with a credit. Asset are on the left side of the accounting equation, and increases to assets are shown on the debit side of the account. Likewise, liabilities are on the right side of equation, and increases to liabilities are shown on the credit side of the account.

Transaction (f): Andy withdrew $ 300 for personal use.

Cash

(a)	$ 30,000	(c)	$ 350
		(d)	$ 500
		(f)	$ 300

Ted Andy, Drawings

(f) $ 300	

分析：
现金减少 300 元，记入帐户的贷方。Ted Andy 的提款帐户用来记录所有者的累计提款额，减少所有者权益，因此提款帐户为借方记录。

Analysis
To decrease the asset account, cash was credited for $ 300. Remember, a separate account, Drawing of Ted Andy, is used to accumulate withdrawals by the owner. Therefore, to decrease owner's equity, the drawing account was debited.

3.2 Use of Revenue and Expense Accounts

Revenue and expense accounts are to accumulated increases and decreases to owner's equity. By having a separate account for each type of revenue and expense, a clear record can be kept.

Also, revenues and expenses can be kept separate from additional investments and withdrawals by the owner. The relationship of these accounts to owner's equity, and the rules of debit and credit are indicated in the following diagram:

All Owner's Equity Accounts

Debit to enter	Credit to enter
Decreases (−)	Increases (+)

All Expense Accounts		All Revenues Accounts	
Debit to enter	Credit to enter	Debit to enter	Credit to enter
Increases (+)	Decreases (−)	Decreases (−)	Increases (+)

Transaction (g): Received $ 3,500 in cash from a client for professional services rendered.

Cash

(a)	$ 30,000	(c)	$ 350
(g)	$ 3,500	(d)	$ 500
		(f)	$ 300

	Professional Fees	
	(g)	$ 3,500

分析：
此項經濟交易增加現金，同時由於收入增加，所有者權益也增加。現金帳戶記入借方，收入帳戶「專業服務收入」記入貸方。「專業服務收入」是一個臨時帳戶，最終導致所有者權益增加。

Analysis
This transaction increased the cash, with an equal increase in owner's equity, because of revenue. The asset account cash was debited and the revenue account professional fees was credited. Professional fees is a temporary account that has the overall effect of increasing owner's equity.

Transaction (h): Paid $ 1,000 for office rent for one month.

	Cash		
(a)	$ 30,000	(c)	$ 350
(g)	$ 3,500	(d)	$ 500
		(f)	$ 300
		(h)	$ 1,000

	Rent Expenses	
(h)	$ 1,000	

分析：
此項經濟交易減少現金，同時費用的增加導致所有者權益減少。借記租金費用 $ 1,000，貸記現金 $ 1,000。租金費用帳戶是一個臨時帳戶，最終導致所有者權益減少。

Analysis
This transaction decreased cash, with an equal decrease in owner's equity because of expense. Rent expense was debited and cash was credited for $ 1,000. Rent expense is a temporary account that has the overall effect of decreasing owner's equity.

Transaction (i): Paid bill for telephone service, $ 75.

	Cash		
(a)	$ 30,000	(c)	$ 350
(g)	$ 3,500	(d)	$ 500
		(f)	$ 300
		(h)	$ 1,000
		(i)	$ 75

	Telephone Expenses
(i)	$ 75

分析:
借記電話費 $ 75，貸記現金 $ 75。

Analysis
This transaction is identical to the previous one. Telephone expense was debited and cash was credit for $ 75.

Chapter 6　Sales

企業銷售商品或提供勞務，並獲得收入。

很多企業的經濟交易涉及增值稅。增值稅是供應商品或勞務時徵收的稅種。

Business exist to sell goods or services, from which they earn income. Income from selling goods or services is usually called sales revenue, turnover, or simple sales. Many business transactions involve sales tax (e.g., VAT in the UK). Sales tax is charged on the supply of goods and services. If you understand the principle behind the tax and how it collected, you will understand the accounting treatment.

Unit 1　Sales Revenue

1.1　Sale of Goods

銷售商品收入滿足下列條件時，予以確認：

·企業已將商品所有權上的主要風險和報酬轉移給購貨方。

·企業沒有保留與所有權相聯繫的繼續管理權，也沒有對已售出的商品實施有效控制。

·收入的金額能夠可靠地計量。

·相關的經濟利益很可能流入企業。

·相關的已發生的成本能夠可靠地計量。

Revenue from the sales of goods should only be recognised when all followings conditions are satisfied.

· The entity has transferred the significant risks and rewards of ownership of the goods to the buyer.

· The entity has no continuing managerial involvement to the degree usually associated with ownership, and no longer has effective control over the goods sold.

· The amount of revenue can be measured reliably.

· It is probable that the economic benefits associated with the transaction will flow to the entity.

· The costs incurred in respect of the transaction can be measured reliably.

The transfer of risks and rewards can only be decided by examining each transaction. Mainly, the transfer occurs at the same time as either the transfer of legal title, or the passing of possession to the buyer – this is what happens when you buy something in a shop.

If significant risks and rewards remain with the seller, then the transaction is not a sale and revenue cannot be recognised. For example, if the receipt of the revenue from a particular sale depends on the buyer receiving revenue from his own sale of the goods.

1.2 Rendering of Services

When the outcome of a transaction involving the rendering of services can be estimated reliably, the associated revenue should be recognised by reference to the stage of completion of the transaction at the balance date. The outcome of a transaction can be estimated reliably when all the following conditions are satisfied.

・收入的金額能夠可靠地計量。

・相關的經濟利益很可能流入企業。

・資產負債表日，經濟業務的完成程度可以可靠地計量。

・相關的已發生和至完工將要發生的成本能夠可靠計量。

・The amount of revenue can be measured reliable.

・It is probable that the economic benefits associated with the transaction will flow to the entity.

・The stage of completion of the transaction at the balance sheet date can be measured reliably.

・The costs incurred for the transaction and the costs to complete the transaction can be measured reliably.

There are various methods of determining the stage of completion of a transaction, but for practical purposes, when services are performed by an indeterminate number of acts over a period of time, revenue should be recognised on a straight-line basis over the period, unless there is evidence for the use of a more appropriate method. If one act is more significance than the others, then the significant act should be carried out before revenue is recognised.

Unit 2　Sales Tax

2.1　Principles of Sales Tax

增值稅是對產品的最終消費者徵收的稅種。

Sales tax is levied on the final consumer of a product. Unless they are the final consumer of the product or service then a business that is registered for sales tax is essentially the collection agent for the relevant authority.

購進貨物的增值稅（進項稅額）和銷售貨物的增值稅（銷項稅額）。

企業在經濟交易中發生不含稅的費用和獲得不含稅的收入。

如果銷項稅額大於進項稅額，企業向稅務機關繳納差額部分。如果進項稅額超過銷項稅額，稅務機關將返還差額部分。

Sales tax is charged on purchases (input tax) and sales (output tax). A business registered for sales tax will effectively pay over the sales tax it has added to its sales and recover the sales tax it has paid on its purchases. To this, the business incurs no sales tax expenses and earns no sales tax income. Therefore, sales tax is excluded from the reported sales and purchases of the business.

Periodically, the business calculates the total amount of sales tax added to sales and the total sales tax added to purchases. If output tax (on sales) exceeds input tax (on purchases), the business pays the excess to the tax authorities. If input tax exceeds output tax, the business is repaid the excess by the tax authorities.

Example 6.1

Table 6.1

	Price net of sales tax $	Sales tax 15% $	Total price $
Manufacturer purchases raw materials and components	40	6	46
Manufacturer sells the completed television to a wholesaler	200	30	230
The manufacturer hands over to tax authorities		24	
Wholesaler purchases television	200	30	230
Wholesaler sells television to a retailer	320	48	368
Wholesaler hands over to tax authorities		18	
Retailer purchases television	320	48	368
Retailer sells television	480	72	552
Retailer hands over to tax authorities		24	
Customer purchases television	480	72	552

Table 6.1 shows the total tax of $ 72 is borne by the ultimate consumer but is collected on behalf of the tax authorities at the different stages in the product's life. If we assume that the sales tax of $ 6 on the initial supplies to the manufacturer is paid by the supplier, the tax authorities would collect the sales tax as follows:

	$
Supplier of materials and components	6
Manufacturer	24
Wholesaler	18
Retailer	24
Total sales tax paid	72

Chapter 6 Sales

增值稅不影響利潤表，除非進項稅額不能抵扣。

So sales tax does not affect the statement of profit or loss, unless it is irrecoverable (see below). But it is simply being collected on behalf of the tax authorities to whom a payment is made.

2.2 Accounting for Sales Tax

採購貨物支付增值稅
借：採購成本（不含稅）
借：增值稅（進項稅額）
貸：應付帳款/現金(含稅)

(1) Sales Tax Paid on Purchases (Input Tax)
Dr   Purchases - excluding sales tax (net cost)
Dr   Sales tax (input sales tax)
Cr   Payables / Cash-including sales tax (gross cost)

· The purchases account does not include sales tax because it is not an expense - it will be recovered.

· The payables account does include sales tax, as the supplier must be paid the full amount due.

銷售貨物徵收增值稅
借：應收帳款（含稅）
貸：收入（不含稅）
貸：增值稅（銷項稅額）

(2) Sales Tax Charged on Sales (Output tax)
Dr   Receivables - including sales tax (gross price)
Cr   Sales - excluding sales tax (net selling price)
Cr   Sales tax (output sales tax)

· The sales account does not include sales tax because it is not income - it will have to be paid to the tax authorities.

· The receivables account does include sales tax, as the customer must pay the full amount due.

支付增值稅
借：增值稅
貸：現金

(3) Payment of Sales (Output) Tax
Dr   Sales tax (amount paid)
Cr   Cash (amount paid)

If output tax exceeds input tax, a payment must be made to the tax authorities.

增值稅返回
借：現金
貸：增值稅

(4) Receipt of Sales (Output) Tax
Dr   Cash (amount received)
Cr   Sales tax (amount received)

If input tax exceeds output tax, there will be a receipt from the tax authorities.

2.3 Irrecoverable Sales Tax

不能抵扣的增值税应计入采购成本。

Some sales tax is irrecoverable. Where sales tax is irrecoverable it must be regarded as part of the cost of the items purchased and included in the statement of profit or loss, or in the statement of financial position as appropriate.

For example, if a business pays $500 for entertaining expenses and suffers irrecoverable input tax of $75 on this amount, the total of $575 paid should be charged to the statement of profit or loss as an expense. Similarly, if a business should capitalise the full amount of $5,400 as a non-current asset in the statement of financial position.

2.4 Sales Tax in Day Books

If a business is registered for sales tax, the sales and purchases day book must include entries to record the tax.

Table 6.2 **Sales Day Book**

Date	Invoice	Customer	Ledger Ref	Gross	Sales tax	Net
				$	$	$
03/07/20×6	0701	Spencer	J1	587.50	87.50	500.00
08/07/20×6	0702	Archie	S5	705.00	105.00	600.00
				1,292.50	192.50	1,100.00

The double entry arising from the sales day book (Table 6.2) will be shown as follows:

借：应收帐款　　$1,292.50

贷：增值税　　$192.50

贷：销售收入　　$1,100

Dr　Receivables ledger control account　　$1,292.50

Cr　Sales tax　　$192.50

Cr　Sales　　$1,100.00

Table 6.3 **Purchase Day Book**

Date	Supplier	Ledger Ref.	Gross	Sales tax	Net
			$	$	$
03/07/20×6	Peggy	Y1	1,762.50	262.50	1,500.00
08/07/20×6	Zena	Z8	352.50	52.50	300.00
			2,115.00	315.00	1,800.00

The double entry arising from the sales day book (Table 6.3) will be shown as:

借：採購成本	$ 1,800	Dr	Purchases	$ 1,800.00
借：增值稅	$ 315	Dr	Sales tax	$ 315.00
貸：應付帳款	$ 2,115	Cr	Payable ledger control account	$ 2,115.00

Chapter 7 Inventory

存貨是企業資產負債表中的一項重要資產。存貨對利潤表也有影響，並直接影響毛利的計算。

Inventory is one of the most important assets in a company's statement of financial position. As we will see, it also affects the statement of profit or loss, having a direct impact on gross profit.

In this chapter, you will be required to consider the impact of the relevant International Accounting Standard on the valuation and presentation of an item in the accounts IAS 2 Inventories.

Unit 1 Accounting for Inventory

期初存貨的成本應計入銷售成本，與採購成本一併構成當年可供出售商品存貨的成本。

期末存貨是期末持有尚未出售的商品存貨，應從銷售成本中扣除。

In order to be able to prepare a set of financial statements, inventory must be accounted for at the end of the period. Opening inventory must be included in cost of sales as these goods are available for sale along with purchases during the year. Closing inventory must be deducted from cost of sales as these goods are held at the period end and have not been sold. Accrual concepts requires that the cost of inventories be charged to the statement of profit or loss in the period in which is sold. If all purchases are sold during the period, then purchases = cost of goods sold, which is charged in statement of profit or loss (i.e., no closing inventories). However, if there are inventories, this means cost of goods sold will not equal to the purchase in the period. Therefore, adjustments below should be made.

Formats in the statement of profit or loss are shown as Table 7.1.

Table 7.1

	$	$
Sales		×
Less: cost of goods sold		
Opening inventories	×	
Purchases	×	

Table 7.1 (Continued)

	×	
Less: Closing inventories	(×)	(×)
Gross profit		×

In the inventory ledger account the opening inventory will be the brought forward balance from the previous period. This must be transferred to the statement of profit or loss ledger account with the following entry:

Dr Profit or loss account (ledger account)
Cr Opening inventory (ledger account)

存貨的期初餘額應轉入本年利潤帳戶：
借：本年利潤
貸：期初存貨

存貨的期末餘額也轉入本年利潤帳戶：
借：期末存貨
貸：本年利潤

The closing inventory is entered into the ledger accounts with the following entry:

Dr Closing inventory (ledger account)
Cr Profit or loss account (ledger account)

本年利潤帳戶記錄了期初存貨餘額和期末存貨餘額。存貨帳戶僅反應期末存貨餘額，該餘額在資產負債表中列示。

Once these entries have been completed, the statement of profit or loss ledger account contains both opening and closing inventory and the inventory ledger account shows the closing inventory for the period to be shown in the statement of financial position.

Example 7.1

We will now see how the ledger accounts for inventory are prepared.

We will look at the ledger accounts at the following times:

(a) 編製試算平衡表前的分類帳戶。

(a) Immediately before extracting a trial balance on 31 December 20×7.

(b) 期末調整後的帳戶，並結帳。

(b) Immediately after the year end adjustments and closing off the ledger accounts.

Solution

(a) Ledger accounts before extracting a trial balance.

Inventory			
20×7	$		$
1 Jan Balance b/f	9,500		

存貨是資產，因此期初存貨在存貨帳戶中為借方餘額。

The inventory is an asset, so is a debit entry in the inventory account.

Purchases

20×7	$		$
Various suppliers	150,000		

Sales Revenue

	$	20×7	$
		Various customers	215,000

The balance of $9,500 in inventory account originated from last year's statement of financial position when it appeared as closing inventory. This figure remains unchanged in the inventory account until the very end of the year when closing inventory on 31 December 20×7 is considered.

The closing inventory figure is not usually provided to us until after we have extracted the trial balance on 31 December 20×7.

The purchases and sales figures have been built up over the year and represent the year's accumulated transactions.

試算平衡表中包括與存貨交易相關的期初存貨、採購成本和銷售收入。

The trial balance will include opening inventory, purchases and sales revenue in respect of the inventory transactions.

(b) Ledger accounts reflecting the closing inventory.

期末存貨價值 $7,500

Closing inventory for accounting purposes has been valued at $7,500.

Step 1:

The statement of profit or loss forms part of the double entry. At the year end the accumulated totals from the sales and purchases accounts must be transferred to it using the following journal entries:

年末，銷售收入總額和採購總成本轉入本年利潤帳戶：

借：銷售收入 $215,000
貸：本年利潤 $215,000
借：本年利潤 $150,000
貸：採購成本 $150,000

Dr Sales revenue $215,000
Cr Profit or loss account $215,000
Dr Profit or loss account $150,000
Cr Purchases $150,000

These transfers are shown in the ledger accounts below.

Step 2:

為了計算銷售成本，期初存貨也應轉入本年利潤。

The opening inventory figure ($9,500) must also be transferred to the statement of profit or loss account in order to arrive at cost of sales.

| 借：本年利潤 $9,500 | Dr Profit or loss account $9,500 |
| 貸：(期初)存貨 $9,500 | Cr (Opening) Inventory $9,500 |

Step 3:

The statement of profit or loss can not be completed (and hence gross profit calculated) until the closing inventory is included.

將期末存貨計入利潤表，並結算毛利。	
借：(期末) 存貨 $7,500	Dr (Closing) Inventory $7,500
貸：本年利潤 $7,500	Cr Profit or loss account $7,500

After summarising and balancing off, the ledger then becomes as follows:

Inventory

20×7	$	20×7	$
1 Jan Bal b/f	9,500	31 Dec P/L	9,500
31 Dec P/L	7,500	31 Dec Bal c/f	7,500
	17,000		17,000
20×8			
1 Jan Bal b/f	7,500		

Purchases

20×7	$	20×7	$
Various dates Payables	150,000	31 Dec Profit / Loss	150,000

Sales Revenue

20×7	$	20×7	$
31 Dec Profit / Loss	215,000	Various dates Receivables	215,000

Profit or Loss Account

20×7	$	20×7	$
31 Dec		31 Dec	
Purchases	150,000	Sales revenue	215,000
Inventory	9,500	Inventory	7,500
Gross profit c/f	63,000		
	222,500		222,500
		Gross profit b/f	63,000

Unit 2　Counting Inventory

期末，企業可以通過存貨實地盤點或永續盤存制獲得企業存貨的持有數量。

The quantity of inventories held at the year end is established by means of a physical count of inventory in an annual counting exercise, or by a 『continuous』 inventory count.

Business trading is a continuous activity, but accounting statements must be drawn up at a particular date. In preparing a statement of financial position it is necessary to 『freeze』 the activity of a business so as to determine its assets and liabilities at a given moment. This includes establishing the quantities of inventories on hand, which can create problems.

如果企業的存貨數量較少，則報告日存貨的數量可以通過實地盤點予以確定。

In simple cases, when a business holds easily counted and relatively small amounts of inventory, quantities of inventories on hand at the reporting date can be determined by physically counting them in an inventory count.

如果企業持有大量的不同種類的存貨，則可以通過連續地存貨記錄確定期末存貨的數量。

In more complicated cases, where a business holds considerable quantities of varied inventory, an alternative approach to establishing quantities is to maintain continuous inventory records. This means that a card is kept for every item of inventory, showing receipts and issues from the stores, and a running total. A few inventory items are counted each day to make sure their record cards are correct – this is called a 『continuous』 count because it is spread out over the year rather than completed in one count at a designated time.

One obstacle is overcome once a business has established how much inventory is on hand. But another of the problems noted in the introduction immediately raises its head. What value should the business place on those inventories?

Unit 3　Valuing Inventory

存貨應當按照成本和可變現淨值孰低法計量。確定存貨的成本，可以採用先進先出法或移動加權平均法。

The value of inventories is calculated at the lower cost and netrealizable value for each separate item or group of items. Cost can be arrived at by using FIFO (first in, first out) or AVCO (weighted average costing).

3.1 The Basic Rule

There are several methods which, in theory, might be used for the valuation of inventory.

· Inventories might be valued at their expected selling price.

· Inventories might be valued at their expected selling price, less any costs still to be incurred in getting them ready for sale and then selling them. This amount is referred to as the net realizable value (NRV) of the inventories.

· Inventories might be valued at their historical cost (i.e., the cost at which they were originally bought).

· Inventories might be valued at the amount it would cost to replace them. This amount is referred to as the current replacement cost of inventories.

·存貨按照估計售價計價。

·存貨按照估計售價減去至完工並銷售估計將要發生的成本進行計價。這個計價金額就是存貨的可變現淨值。

·存貨按照歷史成本計價。

·存貨按照重新採購存貨發生的成本計價。這個金額就是存貨的現行重置成本。

Current replacement costs are not used in the type of accounts dealt with in this syllabus.

The use of selling prices in inventory valuation is ruled out because this would create a profit for the business before the inventory has been sold.

A simple example might help to explain this. A trader buys two items of inventory, each costing $ 100. He can sell them for $ 140 each, but in the accounting period we shall consider, he has only sold one of them. The other is closing inventory in hand.

Since only one item has been sold, you might think it is common sense that profit ought to be $ 40. But if closing inventory is valued at selling price, profit would be $ 80 (i.e., profit would be taken on the closing inventory as well).

This would contradict the accounting concept of **prudence** (i.e., to claim a profit before the item has actually been sold).

The same objection **usually** applies to the use of NRV in inventory valuation. The item purchased for $ 100, requires $ 5 of further expenditure in getting it ready for sale and then selling it (e.g., $ 5 of processing costs and distribution costs). If its expected selling price is $ 140, its NRV is $ (140-5) = $ 135. To value it at $ 135 in the statement of financial position would still be to anticipate a $ 35 profit.

Table 7.2

	$	$
Sales		140
Less: cost of goods sold		
Opening inventories		–
Purchases	200	
	200	
Less: Closing inventories	(140)	(60)
Gross profit		80

存貨應當按照歷史成本進行計價。特殊情況下，謹慎性原則要求採用更低價值對存貨計價時，則不再按照歷史成本計價。

We are left with **historical cost** as the normal basis of inventory valuation. (**The only time when historical cost is not used in the exceptional cases where the prudence concept requires a lower value to be used.**)

Staying with the example above, suppose that the market in this kind of product suddenly slumps and the item's expected selling price is only $ 90. The item's NRV is then $ (90−5) = $ 85 and the business has in effect made a loss of $ 15 ($ 100 − $ 85). The prudence concept requires that losses should be recognized as soon as they are foreseen. This can be achieved by valuing the inventory item in the statement of financial position at its NRV of $ 85.

存貨應當按照成本與可變現淨值孰低法計量。

The argument developed above suggests that the rule to follow is that inventories should be valued at cost, or if lower, net realizable value. The accounting treatment of inventory is governed by an accounting standard, IAS 2 Inventories. IAS 2 states that **inventory should be valued at the lower of cost and net realizable value**.

This is an important rule which you should learn by heart.

3.2 Applying the Basic Valuation Rule

如果企業的存貨很多，應當分別比較每一單項存貨的成本與可變現淨值。

If a business has many inventory items on hand, the comparison of cost and NRV should theoretically be carried out for each item separately. It is not sufficient to compare the total cost of all inventory items with their total NRV. An example will show why.

Suppose a company has four items of inventory on hand at the end of its accounting period. Their cost and NRVs are shown as Table 7.3.

Table 7.3

Inventory item	Cost	NRV	Lower of cost / NRV
	$	$	$
1	27	32	27
2	14	8	8
3	43	55	43
4	29	40	29
	113	135	107

It would be incorrect to compare total costs ($113) with total NRV ($135) and to state inventories at $113 in the statement of financial position. The company can foresee a loss of $6 on item 2 and this should be recognised. If the four items are taken together in total the loss on item 2 is masked by the anticipated profits on the other items. By performing the cost/NRV comparison for each item separately the prudent valuation of $107 can be derived. This is the value which should appear in the statement of financial position.

如果企業的存貨數量繁多，也可分別比較每個存貨類別的可變現淨值和成本總額。

However, for a company with large amounts of inventory this procedure may be impracticable. In this case it is acceptable to group similar items into categories and perform the comparison of cost and NRV category by category, rather than item by item.

Example 7.2

The Table 7.4 shows inventory held at the year end.

Table 7.4

	A	B	C
	$	$	$
Cost	20	9	12
Selling price	30	12	22
Modification cost to enable sale		2	8
Units held	7	2	2
Quantity units	200	150	300

Required: calculate the value of inventory held.

Solution

The answer of question above is shown as Table 7.5.

Table 7.5

Item	Cost	NRV	Valuation	Quantity units	Total value
	$	$	$		$
A	20	23	20	200	4,000
B	9	8	8	150	1,200
C	12	12	12	300	3,600
					8,800

So have we now solved the problem of how a business should value its inventories? It seems that all the business has to do is to choose the lower of cost and net realizable value. This is true as far as it goes, but there is one further problem, perhaps not so easy to foresee: for a given item of inventory, **what was the cost**?

3.3 Determining the Purchase Cost

存貨包括採購的原材料、零部件、已加工完成且待售的產成品以及在產品。

Inventories may be **raw materials** or components bought from suppliers, **finished goods** which have been made by the business but not yet sold, or work in the process of production, but only part-completed (this type of inventory is called **work in progress** or **WIP**). It will simplify matters, however, if we think about the historical cost of purchased raw materials and components, which ought to be their purchase price.

A business may be continually purchasing consignments of a particular component. As each consignment is received from suppliers they are stored in the appropriate bin or on the appropriate shelf or pallet, where they will be mingled with previous consignments. When the storekeeper issues components to production he will simply pull out from the bin the nearest components to hand, which may have arrived in the latest consignment or in an earlier consignment or in several different consignments. Our concern is to devise a pricing technique, a rule of thumb which we can use to attribute a cost to each of the components issued from stores.

Chapter 7 Inventory

先進先出法：假設先入庫的存貨先發出。

移動加權平均法：假設存貨採用加權平均單位成本計量。

「收入」是指收到並入庫的商品存貨；「發出」是指出庫的商品存貨。

There are several techniques which are used in practice.

· **FIFO (first in, first out)**. Using this technique, we assume that components are used in the order in which they are received from suppliers. The components issued are deemed to have formed part of the oldest consignment still unused and are costed accordingly.

· **AVCO (weighted average costing)**. As purchase prices change with each new consignment, the average price of components in the bin is constantly changed. Each component in the bin at any moment is assumed to have been purchased at the average price of all components in the bin at that moment.

If you are preparing **financial accounts** you would normally expect to use FIFO or average costs for the valuation of inventory. You should note furthermore that terms such as AVCO and FIFO refer to **pricing techniques** only. The actual components can be used in any order.

To illustrate the various pricing methods, the following transactions will be used in each case.

Receipts mean goods are received into store and issues represent the issue of goods from store. The problem is to put a valuation on the following:

· The issues of materials.
· The closing inventory.

How would issues and closing inventory be valued using each of the following in turn (Table 7.6)?

Table 7.6   **Transactions during May** 20×7

	Quantity units	Unit Cost	Total cost	Market value per unit on date of transactions
		$	$	$
Opening balance 1 May	100	2.00	200	
Receipts 3 May	400	2.10	840	2.11
Issue 4 May	200			2.11
Receipts 9 May	300	2.12	636	2.15
Issues 11 May	400			2.20
Receipts 18 May	100	2.40	240	2.35
Issues 20 May	100			2.35
Closing balance 31 May	200			2.38
			1,916	

3.4 FIFO (First in, First out)

先進先出法假設存貨依據入庫的先後順序, 既先入庫的存貨先發出。

FIFO assumes that materials are **issued out of inventory in the order in which they were delivered in to inventory** (i.e., issues are priced at the cost of the earliest delivery remaining in inventory).

The cost of issues and closing inventory value in the example, using FIFO, would be as Table 7.7 shows (note that 『OI』 stands for opening inventory).

Table 7.7

Date of issue	Quantity units	Value issued $	Cost of issues $	$
4 May	200	100 OI at $ 2	200	
		100 at $ 2.10	210	
				410
11 May	400	300 at $ 2.10	630	
		100 at $ 2.12	212	
				842
20 May	100	100 at $ 2.12	212	
				212
				1,464
Closing inventory value	200	100 at $ 2.12	212	
		100 at $ 2.40	240	
				452
				1,916

發出存貨成本+期末存貨 = 採購成本+期初存貨 ($ 1,916)

Note that the cost of materials issued plus the value of closing inventory equals the cost of purchases plus the value of opening inventory ($ 1,916).

3.5 AVCO (Weighted Average Cost)

There are various ways in which average costs may be used in pricing inventory issues. The most common (cumulative weighted average pricing) is illustrated below.

The cumulative weighted average pricing method calculates a weighted average price for all units in inventory. Issues are priced at this average cost, and the balance of inventory remaining would have the same unit valuation.

每入庫一批存貨，就計算一個新的加權平均單位成本。這是移動加權平均法的要點。

A new weighted average price is calculated whenever a new delivery of materials into store is received. This is the key feature of cumulative weighted average pricing.

In the example, issue costs and closing inventory values would be shown as Table 7.8.

Table 7.8

Date	Received units	Issued units	Balance units	Total inventory value	Unit cost	Price of issue
				$	$	$
Opening inventory			100	200	2.00	
3 May	400			840	2.10	
			500	1,040	2.08*	
4 May		200		(416)	2.08**	416
			300	624	2.08	
9 May	300			636	2.12	
			600	1,260	2.10*	
11 May		400		(840)	2.10**	840
			200	420	2.10	
18 May	100			240	2.40	
			300	660	2.20*	
20 May		100		(220)	2.20**	220
						1,476
Closing inventory			200	440	2.20	440
						1,916

*：每入庫一批存貨，就計算一個新的存貨單位成本。

**：存貨發出時，以最新的加權平均成本計價。

*: A new unit cost of inventory is calculated whenever a new receipt of materials occurs.

**: Whenever inventories are issued, the unit value of the items issued is the current weighted average cost per unit at the time of the issue.

發出存貨成本+期末存貨=採購成本+期初存貨（$ 1,916）

For this method, the cost of materials issued plus the value of closing inventory equals the cost of purchases plus the value of opening inventory ($ 1,916).

3.6　Inventory Valuations and Profit

不同的存貨計價方法計算的期末存貨和發出存貨成本各不相同。原材料的成本影響生產成本，生產成本最終轉化為銷售成本。因此，採用不同的存貨計價方法，利潤額也是不相同的。

In the previous descriptions of FIFO and AVCO, the example used raw materials as an illustration. Each method of valuation produced different costs both of closing inventories and also of material issues. Since raw material costs affect the cost of production, and the cost of production works through eventually into the cost of sales, it follows that different methods of inventory valuation will provide different profit figures.

Chapter 8 Non-current Assets

Non-current assets can be expensive items and so can have a big impact on a business's financial statements.

非流動資產和流動資產不同，因為它們：

Non-current assets are distinguished from current assets because they:

・可以是有形資產或無形資產；

・could be tangible assets or intangible assets;

・供企業長期使用；

・are bought by the business for use in the long term;

・不以銷售為目的；

・are not normally acquired for resale;

・直接或間接為企業帶來收益；

・are used to generate income directly or indirectly for a business;

・流動性較弱。

・are not normally liquid assets (i.e., not easily and quickly converted into cash without a significant loss in value).

Unit 1 Acquisition of a Non-current Asset

資本性支出和收益性支出

1.1 Capital and Revenue Expenditure

It follows that a business' expenditure may be classified as one of two types:

・capital expenditure;
・revenue expenditure.

Table 8.1 Two Types of Expenditure

Capital expenditure	Revenue expenditure
・Expenditure on the acquisition of non-current assets required for use in the business, not for resale. ・Expenditure on existing non-current assets aimed at an improvement in their earning capacity. ・Expenditure forms part of the cost of non-current assets.	・Expenditure incurred for the purpose of the trade (e.g., purchase of raw materials). ・Expenditure relating to running the business, such as administration costs. ・Expenditure on maintaining the existing earning capacity of non-current assets (e.g., repairs and renewals).

Table 8.1（Continued）

Capital expenditure	Revenue expenditure
· Capital expenditure is long-term in nature as the business intends to receive the benefits of the expenditure over a long period of time. · Capital expenditure is not charged as an expense in the statement of profit or loss.	· Revenue expenditure relates to the current account period and is used to generate revenue in the business. · Revenue expenditure must be expense in the statement of profit or loss.
資本性支出	收益性支出
·購置供企業使用，不以銷售為目的非流動資產發生的支出 ·用於提高現有固定資產獲利能力的支出 ·支出計入非流動資產的成本	·以交易為目的發生的支出，如採購原材料 ·和經濟業務營運相關的支出，如管理費用 ·用於維持現有固定資產獲利能力的開支，如修理費和零配件更新
·資本性支出具有長期性，企業預期將在較長的時期內獲得經濟利益 ·資本性支出不能作為一項費用計入利潤表	·收益性支出是在一定會計期間為獲得收入而產生的支出 ·收益性支出作為一項費用計入利潤表

1.2 Example Capital and Revenue Expenditure

A business purchase a building for ＄50,000. It then adds an extension to the building at cost of ＄20,000. The building needs to have a few broken windows mended, some missing roof tiles replaced and its floors polished. These cleaning and maintenance jobs cost ＄1,900.

初始購置成本（＄50,000）和擴建成本（＄20,000）屬於資本性支出。

In this example, the original purchase (＄50,000) and the cost of the extension (＄20,000) are capital expenditure, because they are incurred to acquire and then improve a non-current asset and so can be capitalised as part of it. The other costs of ＄1,900 are revenue expenditure, because these merely maintain the building and thus the 『earnings capacity』 of the building.

其他開支＄1,900屬於收益性支出。

非流動資產備查簿

1.3 Non-current Asset Registers

A non-current asset register is maintained in order to control non-current assets and keep track of what is owned and where it is kept.

It forms part of the internal control system of an organization, to make sure that the information about non-current assets reported in the nominal ledger is accurate and correct.

The non-current asset register lists out all the details of each non-current asset that is owned by the business.

Details held on such a register may include:
- cost;
- date of purchase;
- description of asset;
- internal reference number;
- manufacturer's serial number;
- location of asset;
- deprecation method;
- expected useful life;
- carrying amount (net book value).

初始計量

1.4 Initial Measurement

The cost of a non-current asset is any amount incurred to acquire the asset and bring it into working condition.

Table 8.2

Includes	Excludes
Capital expenditure such as: • Purchase price (including any import duties paid, but excluding any trade discount and sales tax paid) • Delivery and handling costs • Professional fees • Trials and tests • Subsequent expenditure which enhances the asset • Staff costs arising directly from the construction of the asset	資本性支出： • 買價（包括支付的進口關稅，不包括商業折扣和支付的增值稅） • 運輸費和裝卸費 • 專業人員服務費 • 測試費 • 提升資產性能的後續支出 • 建造資產發生的直接人工費
包括	不包括
Revenue expenditure such as: • Repairs • Renewals • Repainting • Administration • General overheads • Training costs • Wastage	收益性支出： • 修理 • 更換配件 • 重新粉刷 • 管理費 • 間接費用 • 培訓費 • 損耗

借：非流動資產——成本
貸：現金（應付款項）

The correct double entry to record the purchase is:
Dr   Non-current asset - cost $ ×
Cr   Cash (or sundry payable) $ ×

A separate cost account should be kept for each category of non-current asset (e.g., motor vehicles, fixtures and fittings).

1.5 Subsequent Expenditure

後續支出

如果非流動資產的後續支出能增加資產帶來的經濟利益，則該支出計入非流動資產成本。

Subsequent expenditure on the non-current asset can only be recorded as part of the cost (or capitalized), if it enhances the benefits of the asset.

The important point here is whether subsequent expenditure on an asset improves the condition of the asset beyond the previous performance.

An example of subsequent expenditure which meets this criterion, and so can be capitalized, is an extension to a shop building which provides extra selling space.

修理費是不滿足資本化條件的後續支出。
因此，修理支出應該確認為一項費用，並計入利潤表。

An example of subsequent expenditure which does not meet this criterion is repair work. Any repair and maintenance costs merely maintain or restore value, so such cost must be recognised as an expense and debited to the statement of profit or loss.

Unit 2　Depreciation

Deprecation accounting is governed by **IAS 16 Property, Plant and Equipment**. IAS 16 defines depreciation as 「the measure of the cost or revalued amount of the economic benefits of the tangible non-current asset that has been consumed during the period」.

折舊就是將非流動資產的成本在其使用壽命內進行分攤，使資產的成本和相應的收入相配比。

Depreciation is a means of spreading the cost of a non-current asset over its useful life, in order to match the cost of the asset with the profits it earns for the business.

Depreciation must also be matched to the pattern of use of the asset. This must be regularly reviewed and may be changed if the method no longer matches the usage of the asset.

This is achieved by recording a depreciation charge each year, the effect of which is twofold (『the dual effect』).

· Reduce the statement of financial position value of the non-current asset by accumulated depreciation to reflect the wearing out.

· Depreciation as an expense in the statement of profit or loss to match to the revenue generated by the non-current asset.

Depreciation may arise from:

· use;

· expected physical wear and tear;

· passing of time (e.g., a ten-year lease on a property);

· obsolescence through technology and market changes (e.g., plant and machinery of a specialized nature);

· depletion (e.g., the extraction of a mineral from a quarry).

The purpose of depreciation is not to show the asset at its current value in the statement of financial position, nor is it intended to provided a fund for the replacement of the asset. It is simply a method of allocating the cost of the asset over the periods estimated to benefit from its use (the useful life).

Land normally has an unlimited life and so does not required depreciation, but buildings should be depreciated.

Depreciation of an asset begins when it is available for use.

2.1 Methods of Calculating Depreciation

2.1.1 Straight-line Method

$$\text{Annual depreciation charge} = \frac{\text{Cost of asset} - \text{Residual value}}{\text{Expected useful life}}$$

$$年折舊額 = \frac{資產原值 - 預計淨殘值}{預計使用年限}$$

Residual value: the estimated disposal value of the asset at the end of its useful life.

Useful life: the estimated number of years during which the business will use the asset. The useful life does not necessarily equal the physical life of the assets. For example many businesses use a three-year useful life for computers.

This does not mean that a computer can no longer be used after three years. It means that the business is likely to replace the computer after three years due to technological advancement.

The straight line method results in the same charge every year and is used wherever the pattern of usage of an asset is consistent throughout its life.

Buildings are commonly depreciation using this method because businesses will commonly get the same usage out of a building every year.

Example 8.1

A non-current asset costing $ 60,000 has an estimated life of 5 years and a residual value of $ 2,400. The annual depreciation charge using the straight line method would be:

$$\frac{\$ (60,000-2,400)}{5 \text{ years}} = \$ 11,520 \text{ per annum}$$

Alternatively it can simple be given as a simple percentage of cost.

$$\frac{1-4\%}{5 \text{ years}} = 19.2\%$$

This is the same as saying 「the depreciation charge per annum is 19.2% of cost' (i.e., 19.2% of $ 60,000 = $ 11,520.

2.1.2 Reducing Balance Method

餘額遞減法

折舊 = 折舊率×帳面金額

帳面金額：非流動資產的原值減資產的累計折舊

Depreciation charge = X%×Carrying amount

Carrying amount: original cost of the non-current asset less accumulated depreciation on the asset to date.

The reducing balance method results in a constantly reducing depreciation charge throughout the life of the asset. This is used to reflect the expectation that the asset will be used less and less as it ages. This is a common method of depreciation for machinery, where it is expected that they will provide less service to the business as they age because of the increased need to service / repair them as their mileage increases.

Example 8.2

A business purchases a lorry at a cost of $ 17,000. Its estimated residual value is $ 2,000. It is expected to last for five years.

Work out the depreciation to be charged each year under the reducing balance method (using a rate of 35%).

Solution

Table 8.3 shows the solution of Example of 8.2.

Table 8.3

Year	Depreciation charge % × CV	Depreciation charge
1	35% × $ 17,000	$ 5,950
2	35% × $ (17,000− $ 5,950)	$ 3,868
3	35% × $ (11,050− $ 3,868)	$ 2,514
4	35% × $ (7,182− $ 2,514)	$ 1,634
5	$ 4,668− $ 1,634− $ 2,000	$ 1,034

會計估計變更

2.2 Changing Estimates

Businesses should apply the same rates and methods of depreciation consistently throughout the life of their business. However, if they believe that their estimates of useful life or residual value are inappropriate they are permitted to change them with no further recourse. In order to do this you simply work out the new depreciation charge of the asset based on the revised estimate of useful life or residual value.

Example 8.3

J Co. purchased a non-current asset for $ 100,000 on 1 January 20×7 and started depreciating it over five years. Residual value was taken as $ 10,000.

On 1 January 20×8 a review of asset lives was undertaken and the remaining useful life of the asset was estimated at eight years. Residual value was estimated to be nil.

Calculate the depreciation charge for the year ended 31 December 20×8 and subsequent years.

Solution

Initial depreciation charge = ($ 100,000− $ 10,000) / 5 years = $ 18,000 per annum

On 1 Jan 20×8 the asset would have accumulated one year's worth of depreciation. Its carrying amount would therefore be $100,000 - $18,000 = $82,000.

At this point the asset is estimated to have a remaining useful life of 8 years and $nil residual value. From now on the depreciation charge will be $82,000 / 8 years = $10,250 pa.

2.3 Accounting for Depreciation

The ledger accounting entries for depreciation are as follows.

Dr　Depreciation expense (SPL)
Cr　Accumulated depreciation (SOFP)

With the depreciation charge for the period.

· The depreciation expense account is a profit or loss account and therefore it is not accumulated.

· The accumulated depreciation account is a statement of financial position account and as the name suggests is accumulated (i.e., reflects all depreciation to date).

· The non-current asset accounts are unaffected by depreciation. Non-current assets are recorded in these accounts at cost.

· On the statement of financial position it is shown as a reduction against the cost of non-current assets to derive the carrying amount of the non-current assets.

	$
Cost	X
Accumulated depreciation	(X)
Carrying amount	X

Unit 3　Disposal of Non-current Asset

3.1 The Disposal of T-account

When a tangible non-current asset is sold, there is likely to be a profit or loss on disposal. This is the difference between the net sales price of the asset and its carrying amount at the time of disposal. Net sales price is the price minus any costs of making the sale.

銷售淨價>處置日資產的帳面價值=>利得	Net sales price>Carrying amount at disposal date=>Profit
銷售淨價<處置日資產的帳面價值=>損失	Net sales price<Carrying amount at disposal date=>Loss
銷售淨價=處置日資產的帳面價值=>不產生利得或損失	Net sales price=Carrying amount at disposal date=>Neither profit or loss
	Note：a disposals T-account is required when recording the disposal of a non-current asset.
	This is a three-step process：
註銷非流動資產原值帳戶	· Remove the original cost of the non-current asset from the 『non-current asset』 account.
	Dr　Disposal of non-current asset
	Cr　Non-current assets account－cost
註銷累計折舊帳戶	· Remove accumulated depreciation on the non-current asset from the 『accumulated depreciation』 account.
	Dr　Accumulated depreciation account
	Cr　Disposal of non-current asset
記錄處置收入	· Record the proceeds.
	Dr　Cash account
	Cr　Disposal of non-current asset
處置帳戶的餘額即為處置利得或損失。	The balance on the disposals T-account is the profit or loss on disposal.

Disposal

Original cost	X	Accumulated depreciation	X
		proceeds	X
Profit on disposal	B	Loss on disposal	B
	X		X

The profits or losses are reported in the income and expenses part of the statement of profit or loss of the business, after gross profit. They are commonly referred to as 『profit on disposal of non-current assets』 or 『loss on disposal』.

3.2 Disposal Through a Part Exchange Agreement

以舊換新處置資產

Exchange and part exchange of assets occurs frequently for items of non-current assets. A part exchange agreement arises where an old asset is provided in part payment for a new one, the balance of the new asset being paid in cash.

There is a four-step process:

註銷非流動資產帳戶

· Remove the original cost of the non-current asset from the 『non-current asset』 account.

Dr　Disposal of non-current asset
Cr　Non-current assets account – cost

註銷累計折舊帳戶

· Remove accumulated depreciation on the non-current asset from the 『accumulated depreciation』 account.

Dr　Accumulated depreciation account
Cr　Disposal of non-current asset

The first two steps are identical, however the last two step are as follows:

將舊資產抵價金額作為收益登記入帳。

· Record the part exchange allowance as proceeds.

Dr　Non-current asset account – cost
　　　(= part of cost of new asset)
Cr　Disposal of non-current asset
　　　(= sale proceeds of old asset)

記錄新購資產現金支付的金額。

· Record the cash paid for the new asset.

Dr　Non-current asset account – cost
Cr　Cash account

Again, the balance on the disposals T-account is the profit or loss on disposal:

Disposal

Original cost	X	Accumulated depreciation	X
		proceeds	X
Profit on disposal	<u>B</u>	Loss on disposal	<u>B</u>
	<u>X</u>		<u>X</u>

Chapter 8　Non-current Assets

Unit 4　Revaluation of Non-current Assets

IAS 16 允許企業按照公允價值重估非流動資產。

IAS 16 allows entities to revalue non-current assets to fair value. Some non-current assets, such as land and buildings may rise in value over time.

企業可以選擇在資產負債表中反應資產的現時價值。

Companies (rather than sole traders and partnerships) may choose to reflect the current value of the asset in their statement of financial position. This is known as revaluing the asset.

4.1　Accounting Entries

資產帳面金額與重估金額之間的差額記入所有者權益的重估增值帳戶。

The difference between the carrying amount of the asset and the revalued amount (normally a gain) is recorded in a revaluation surplus account in the equity or capital section of the statement of financial position.

重估增值＝重估價值-帳面金額

Revaluation surplus＝Revalued amount-carrying amount

For a non-depreciated asset:
Dr　Non-current asset-cost (revaluation surplus)
Cr　Revaluation surplus (revaluation surplus)

For a depreciation asset:
Dr　Non-current asset-cost
Dr　Accumulated depreciation (depreciation to date)
Cr　Revaluation surplus (Revaluation gain)

重估利得屬於沒有實現的利得，不能計入利潤表。

The gain on revaluation cannot be recorded in the statement of profit or loss because it is unrealized. Think about owning a house; the value of the house may appreciate in value over time. You can't go out and spend that increase in value on a new car though because it is not a real gain to you; it only becomes real when you sell the house and receive the benefit from that sale.

Until that point it is a hypothetical gain (i.e., how much you would gain if you sold it at that point in time). Don't forget, the house could fall in value again by the time you sell it leaving you to find other ways to fund that car you just agreed to buy!

IAS 1 要求在綜合收益表中單獨披露當年的重估利得，作為「其他綜合收益」項目。

這個金額與以前年度的重估增值相加，得到所有者權益變動表和資產負債表的累計重估增值金額。

As this increase in value represents an unrealized gain we cannot record it as part of the profit earned during the year. IAS 1 requires that revaluation gain for the year is disclosed separately from profit on the face of the statement of profit and loss, and other comprehensive income as an item of 『other comprehensive income』. This amount is added to any earlier revaluation from a previous accounting period to arrive at a cumulative revaluation surplus in the statement of changes in equity and statement of financial position.

Example 8.4

Ira Vann runs a kilt-making business in London. It has run the business for many years from a building which originally cost $350,000 and on which $140,000 accumulated depreciation has been charged to date. Ira Vann wishes to revalue the building to $800,000.

How is this reflected in the financial statements?

Solution

The current balances in the accounts are:

Building cost $350,000

Accumulated depreciation $140,000

· The building asset account needs to be raised by $450,000 to $800,000

· On revaluation, the accumulated depreciation account is cleared out.

Therefore the double entry required is:

Dr	Non-current asset (building)	$450,000
Dr	Accumulated depreciation	$140,000
Cr	Revaluation surplus	$590,000

重估金額 800,000 美元與重估前資產帳面價值 210,000 美元的差額，即重估增值 590,000 美元。

The gain of $590,000 reflects the difference between the carrying amount pre-revaluation of $210,000 and the revaluated amount of $800,000.

Extract from the statement of profit or loss and other comprehensive income. Other comprehensive income is item that will not be reclassified in subsequent accounting periods: Gain on property revaluation in year $590,000.

Asset (Building)

	$		$
Balance b/f	350,000	Balance c/f	800,000
Revaluation surplus	450,000		
	800,000		800,000
Balance b/f	800,000		

Accumulated Depreciation (Building)

	$		$
Revaluation surplus	140,000	Balance b/f	140,000
	140,000		140,000

Revaluation Surplus

	$		$
		Non-current asset (building)	450,000
Balance c/f	590,000	Accumulated depreciation (building)	140,000
	590,000		590,000
		Balance b/f	590,000

4.2 Excess Depreciation

非流動資產重估後，折舊根據重估金額計提。

計提的折舊
=重估金額/剩餘使用壽命

When a non-current asset is revalued, depreciation is charged on the revalued amount.

$$\text{Depreciation charged} = \frac{\text{Revalued amount}}{\text{Remaining useful life}}$$

Example 8.5

When Perry P Louis commenced trading as a car hire dealer on 1 January 20×1, he purchased a building at a cost of $ 300,000. This would be depreciated by the straight line method to a nil residual value over 30 years. After five years of trading, on 1 January 20×6 Perry P Louis decides that the building is now worth $ 750,000.

Before the revaluation, the annual depreciation charge is $ 10,000 ($ 300,000 / 30 years) per annum on the building. This charge is made in each of the first five years of the asset's life.

After the revaluation, depreciation will be charged on the building at a new rate of:

$ 750,000 / 25 years = $ 30,000 per annum

Dr	Depreciation expense	$ 30,000
Cr	Accumulated depreciation	$ 30,000

To record the new annual depreciation charge.

The difference between the new depreciation charge based on the revalued carrying amount and the old depreciation charge based on the original cost of the asset is known as the 『excess depreciation』.

按照重估價值計提的折舊和按照資產原值計提的折舊之間的差額，就是超額折舊。

Applying this to the example above gives the following.

Excess depreciation = $ 30,000 - $ 10,000 = $ 20,000

IAS 16 allows entities to transfer an amount equal to the excess depreciation from the revaluation surplus to retained earnings in the equity section of the statement of financial position, if entities wish to do so.

如果企業有意願，IAS16 允許企業將超額折舊的金額從重估增值轉入留存收益。

An amount of $ 20,000, representing the excess depreciation, can be transferred each year from the revaluation surplus to retained earnings.

Dr	Revaluation surplus	$ 20,000
Cr	Retained earnings	$ 20,000

To record the transfer of the excess depreciation.

重估增值和留存收益帳戶間金額的轉移反應在所有者權益變動表中。

This transfer between the revaluation surplus to retained earnings is shown in the financial statements in the statement of changes in equity.

Chapter 9 Intangible Non-current Assets

最重要的無形資產是企業的研究和開發支出。

Intangible non-current assets are long-term assets that have a value to the business because they have been paid for, but which do not have any physical substance. The most significant of such intangible assets are research and development (R & D) costs. When R & D is a large item of cost its accounting treatment may have a significant influence on the profit of a business and its statement of financial position valuation. Because this attempts have been made to standardize the treatment, and these are discussed in this chapter.

Unit 1 Intangible Assets

1.1 Types of Intangible Assets

『Intangible』 assets means assets that literally cannot be touched, as opposed to assets (such as plant and machinery) which have a physical existence.

無形資產的特徵：

· 企業控制的資源，預期會給企業帶來未來經濟利益。

· 沒有實物形態。

· 具有可辨認性。

In particular, the key characteristics of an intangible non-current asset are as follows:

· It is a resource controlled by the entity from which the entity expects to derive future economic benefits.

· It lacks physical substance.

· It is identifiable and separately distinguishable from goodwill.

無形資產的確認條件：

· 符合無形資產的定義。

· 無形資產的成本能夠可靠地計量。

· 與無形資產相關的未來經濟利益很可能流入企業。

The basic principle of recognition of an intangible asset in the financial statements is that:

· It should meet the definition of an intangible asset.

· The cost of the asset can be reliably measured.

· That it is probable that future economic benefits will be received by the entity from the asset.

Examples of intangible assets include the following:

- licences;
- patents;
- brands;
- trademarks;
- copyrights;
- franchises.

1.2 Accounting Treatment

Intangible assets can be purchased or may be created within a business without any expenditure being incurred, it means internally generated (e.g., brands). Purchased intangible assets are usually capitalised in the accounts. Generally, internally generated assets may not be capitalsed.

外購無形資產應資本化，內部產生的無形資產則不一定資本化。

If an intangible assets has a finite useful life, it should then be amortised over that life. Amortisation is intended to write off the assets over its useful life (under the accrual concept).

如果無形資產的使用壽命有限，則在使用壽命內進行價值攤銷。

Example 9.1

A business buys a patent for $ 60,000. It expects to use the patent for the next ten years, after which it will be valueless. Amortisation is calculated in the same way as for tangible assets:

$$\frac{\text{Cost} - \text{Residual value}}{\text{Estimated useful life}}$$

In this case, amortisation will be $ 6,000 (60,000/10) per annum.

The double entry treatment for the amortization calculated above will be:

Dr	Amortisation account (statement of profit or loss)	$ 6,000
Cr	Accumul atedamortisation (statement of financial position)	$ 6,000

Unit 2 Research and Development Costs

研發支出

2.1 Research and Development

研究是指為了獲取新的技術和知識等進行的有計劃的調查。

Research can be defined as original and planed investigation undertaken with the prospect of gaining new scientific or technical knowledge and understanding.

Chapter 9 Intangible Non-current Assets   79

開發是指在進行商業性生產和使用前，將研究成果或其他知識應用於某項設計計劃，生產出新的或實質性改進的材料、裝置、產品、工序、服務體系等。

Development can be defined as the application of search findings or other knowledge to a plan of design for the production of new or substantially improved materials, devices, products, processes, systems of services before the start of commercial production or use.

2.2 Components of Research and Development Costs

Research and development costs will included all costs that are directly attributable to research and development activities, or that can be allocated on a reasonable basis.

The standard lists the costs which may be included in R & D, where applicable (note that selling costs are excluded).

· **Salaries, wages** and other employment related costs of personnel engaged in R & D activities.

· **Depreciation** of property, plant and equipment to the extent that these assets are used for R & D activities.

· **Overhead costs**, other than general administrative costs, related to R & D activities; these cost are allocated on bases similar to those used in allocating overhead costs to inventories.

· **Other costs**, such as the amortisation of patents and licences, to the extent that these assets are used for R & D activities.

2.3 Accounting Treatment of Research and Development

Where a company undertakes research and development, expenditure is being incurred with the intention of producing future benefits.

The accounting issue is whether these costs should be expensed to the statement of financial position to match to future benefit arising.

(1) Research

研究階段的支出在發生時應當計入當期損益，體現了謹慎性原則。

All research expenditure should be written off to the statement of profit or loss as it is incurred, this is in compliance with the prudence concept.

Research expenditure does not directly lead to future benefits and therefore it is not possible to follow the matching concept.

Any capital expenditure on research equipment should be capitalised and depreciated as normal.

(2) Development

Development expenditure must be capitalised as an intangible asset provided that certain criteria are met:

· separate project;

· expenditure identifiable and reliably measured;

· commercially viable;

· technically feasible;

· overall profitable;

· resources available to complete.

If the above criteria are not met, development expenditure must be written off to the statement of profit or loss as it is incurred.

Once expenditure has been treated as an expense, it cannot be reinstated as an asset.

Example 9.2

Merlot Co. is engaged in a number of research and development projects. Table 9.1 shows three project.

Table 9.1

Project A	A project to investigate the properties of a chemical compound.
Project B	A project to develop a new process which will save time in the production of widgets. This project was started on 1 January 20×5 and met the capitalisation criteria on 31 August 20×5.
Project C	A development projected which was completed on 30 June 20×5. Related costs in the statement of financial position at the start of the year were $290,000. Production and sales of the new product commenced on 1 September and are expected to last 36 months.

Costs for the year ended 31 December 20×5 are shown as Table 9.2.

Table 9.2

		$
Project A		34,000
Project B	Costs to 31 August	78,870
Project B	Costs from 31 August	27,800
Project C	Cost to 30 June	19,800

What amount is expensed to the statement of profit or loss and other comprehensive income of Merlot Co. in respect of these projects in the year ended 31 December 20×5?

Solution

·A 是研究項目，所有的費用在發生時計入綜合收益表。

· Project A is a research project and all costs should be written off to the statement of profit or loss and other comprehensive income as incurred.

·B 是開發項目。開發支出滿足資本化條件的應該資本化。資本化之前發生的費用不能作為資產轉回。

· Project B is a development project. Costs can only be capitalised once the capitalisation criteria are met. Those costs incurred before the case cannot be reinstated as an asset.

·C 是開發項目，開發支出應資本化。該無形資產在新產品銷售期 36 個月之內攤銷。當年的攤銷期限為 4 個月。

· Project C is s development project which has resulted in capitalised expenditure. This asset must be amortised over the 36 months of sales of the product. Amortisation for the current year should be 4 months (1 September to 31 December 20×5).

Workings:

	$
Project A	34,000
Project B	78,870
Project C ($190,000 + $19,800) × 4/36	34,422
	147,292

2.4 Amortisation of Development Costs

Once capitalised as an asset, development costs must be amortised and recognised as an expense to match the costs with the related revenue or cost savings. This must be done on a systematic basis, to reflect the pattern in which the related economic benefits are recognised.

It is unlikely to be possible to match exactly the economic benefits obtained with the costs which are held as an asset simple because of the nature of development activities. The entity should consider either:

· The revenue or other benefits from the sale/use of the product/process.

· The period of time over which the product/process is expected to be sold/used.

如果無形資產的使用壽命不確定，則不進行攤銷，但應進行減值測試。

If the pattern cannot be determined reliably, the straight-line method should be used. The amortisation will begin when the asset is available for use. If the intangible asset is considered to have an indefinite useful life, it should not be amortised but should be subjected to an annual impairment review.

開發費用的減值

2.5 Impairment of Development Costs

As with all assets, impairment (fall in value of an asset) is a possibility, but is perhaps more likely with development costs, when the asset is linked with success of the development.

The development costs should be written down to the extent that the unamortised balance (taken together with further development costs, related production costs, and selling and administrative costs directly incurred in marketing the product) is no longer likely to be recovered from the expected future economic benefit.

Chapter 10　Accruals and Prepayments

權責發生制是重要的會計概念。

權責發生制是指收入和費用配比，計入所屬當期的利潤表，不論款項是否實際收付。

收入實現相關的費用應與收入相配比。

The accruals concept is identified as an important accounting concept by IAS 1 *Presentation of Financial Statements*.

The concept is that income and expenses should be matched together and dealt with in the statement of profit or loss for the period to which they relate, regardless of the period in which the cash was actually received or paid.

Therefore all of the expense involved in making the sales for a period should be matched with the sales income and dealt with in the period in which the sales themselves are accounted for.

Therefore, profit is:

Income earned	×
Expenditure incurred	(×)
Profit	×

Unit 1　Accrued Expenditure and Prepaid Expenditure

1.1　Accrued Expenditure

應計項目是指發生在當期，但期末尚未支付的費用。

An accrual is an expense that has been incurred during the accounting period but has not been paid at the end of the accounting period.

This means that the accountant must ensure that the expense accounts include not only those items that have been paid for during the period, but also any outstanding amounts incurred.

It is necessary to record the extra expense relevant to the year and create a corresponding statement of financial position liability (called an accrual).

應計項目的會計分錄如下：
借：費用

The journal entry to record a period of accruals is:

Dr　Expense account (SPL)

贷：应计费用

应计项目会导致利润表利润减少。

Cr　Accrual（SOFP）

An accrual will therefore reduce profit in the statement of profit or loss.

Example 10.1

Liza Dolittle has committed to pay a monthly rental of $1,000. During the year of 2015, Liza Dolittle has paid 10 months rental. As at 31 December 2015, he need to accrued the 2 months outstanding under the accrual concept.

Table 10.1

2015											
Jan	Feb	Mar	Apr	May	Jun	Jul	Aug	Sep	Oct	Nov	Dec
$1,000	$1,000	$1,000	$1,000	$1,000	$1,000	$1,000	$1,000	$1,000	$1,000	$1,000	$1,000

(a) Record the cash payments at 31 Decembev 2015.

Rental

			$	
31 Dec	Bank		10,000	

2015年12月31日，计入利润表的租金费用为$12,000。

(b) As at 31 December 2015, the total expense charged to the statement of profit or loss in respect of rent should be $12,000.

Rental

			$				$
31 Dec	Bank		10,000	31 Dec	Profit or loss		12,000

尚未支付的租金是资产负债表中的一项流动负债。

(c) The rental owed will be reported as a current liabilities in the statement of financial position.

Rental

			$				$
31 Dec	Bank		10,000	31 Dec	Profit or loss a/c		12,000
	Accrual c/f		2,000				
			12,000				12,000
				2016			
				1 Jan	Accrual b/f		2,000

1.2 Prepaid Expenditure

預付費用是指當期支付，但歸屬於下一會計期間的費用。

A prepayment is an item of expense that has been paid during the current accounting period but related to the next accounting period.

The prepaid expenses should not appear in the statement of profit or loss, but be recognize in next accounting period.

It is necessary to remove that part of the expense which is not relevant to this year and create a corresponding statement of financial position asset (called a prepayment).

預付費用的會計分錄如下：
借：預付費用
貸：費用

The journal entry to record a prepayment is:
Dr   Prepayment (SOFP)
Cr   Expense Account (SPL)

預付費用增加利潤表利潤。

A prepayment will therefore increase profit in the statement of profit or loss.

Example 10.2

Tubby Wadlow pays the insurance charge for the business in advance. His payments during 2014 for this insurance are shown in Table 10.2.

Table 10.2

1 January	(for three months to 31 March 2014)	$ 600
31 March	(for six months to 30 September 2014)	$ 1,200
30 October	(for six months to 31 March 2015)	$ 1,200

Calculate the insurance expense for the year ended 31 December 2014 and write up the insurance ledger account.

Table 10.3

2014													2015		
Jan	Feb	Mar	Apr	May	Jun	Jul	Aug	Sep	Oct	Nov	Dec	Jan	Feb	Mar
$ 600			$ 1,200						$ 1,200					

(a) Enter the cash payments into the insurance account.

Insurance

2014			$	2014		$
1 Jan	Bank		600			
31 Mar	Bank		1,200			
30 Oct	Bank		1,200			

(b) Charge the expense for the period end to profit or loss account.

Insurance

2014			$	2014		$
1 Jan	Bank		600	31 Dec	Profit or loss a/c	2,400
31 Mar	Bank		1,200			
30 Oct	Bank		1,200			

預付保險費在資產負債表中是一項流動資產。

(c) Prepaid insurance creates a current asset to be included on the statement of financial position.

Insurance

2014			$	2014		$
1 Jan	Bank		600	31 Dec	Profit or loss a/c	2,400
31 Mar	Bank		1,200	31 Dec	Prepayments c/f	600
30 Oct	Bank		1,200			
			3,000			3,000
2015						
1 Jan	Prepayments b/f		600			

Unit 2　Accrued Income and Prepaid Income

2.1　Accrued Income

應計收入是指歸屬於當期，但期末尚未收到的收入。

Accrued income arises where income has been earned in the accounting period but has not yet been received.

It is necessary to record the extra income in the statement of profit or loss and create a corresponding asset in the statement of financial position (called accrued income).

應計收入的會計分錄如下：
借：應計收入
貸：收入

The journal entry to record an accrued income is:
Dr　Accrued income (SOFP)
Cr　Income (SPL)

應計收入增加利潤表利潤。

An accrued income will therefore increase overall profit in the statement of profit or loss

Example 10.3

A business earns bank interest income of $300 per month. $3,000 bank interest income has been received in the year to 31 December 2015.

Write up bank interest income account for the year ended 31 December 2015.

記錄當年的現金收入　　(a) Record the receipts during the year.

Bank Interest Income

2015			2015		$
				Bank	3,000

計入利潤表的利息總收入為 $ 3,600。

(b) The total amount credited to the statement of profit or loss in respect of interest should be $ 3,600 ($ 300 × 12).

Bank Interest Income

2015		$	2015		$
31 Dec	Profit or loss a/c	3,600		Bank	3,000

會計期末，尚未收到的應計收入為 $ 600。

(c) At the end of the period, accrued income asset is the $ 600 that has not yet been received.

Bank Interest Income

2015		$	2015		$
31 Dec	Profit or loss a/c	3,600		Bank	3,000
		————	31 Dec	Accrued income c/f	600
		3,600			3,600
2016					
1 Jan	Accrued income b/f	600			

2.2 Prepaid Income

預收收入是當期收到，歸屬於下一會計期間的收入。

Prepaid income arises where income has been received in the accounting period but which relates to the next accounting period.

It is necessary to remove the income not relating to the year from the statement of profit or loss and create a corresponding liability in the statement of financial position (called prepaid income).

預收收入的會計分錄如下：
借：收入
貸：預收收入

The journal entry to record an accrued income is:
Dr　Income (SPL)
Cr　Prepaid income (SOFP)

預收收入導致收入減少，因此導致利潤表利潤減少。	Prepaid income reduced income on the statement of profit or loss and hence reduced overall profit too.

Example 10.4

A business rents out a property at an income of $4,000 per month. $64,000 has been received in the year ended 31 December 2015.

Write up rental income account for the year ended 31 December 2015.

記錄當年的現金收入。	(a) Record the receipts during the year as follows.

Rental Income

2015			2015		$
				Bank	64,000

計入利潤表的租金總收入為 $48,000。	(b) The total amount credited to the statement of profit or loss in respect of rental income should be $48,000 ($4,000 × 12).

Rental Income

2015		$	2015		$
31 Dec	Profit or loss a/c	48,000		Bank	64,000

會計期末，歸屬於下一年度的預收收入為 $16,000。	(c) At the end of the period, prepaid income liability is the $16,000 that has been received in respect of next year.

Rental Income

2015		$	2015		$
31 Dec	Profit or loss a/c	48,000		Bank	64,000
31 Dec	Prepaid income c/f	16,000			
		64,000			64,000
			2016		
			1 Jan	Prepaid income b/f	16,000

Chapter 11 Provision and Contingencies

It is important to consider the timing of settlement of a liability so that you can ensure you allocate it to the correct category of the statement of financial position.

負債是過去交易或事項形成的，導致經濟利益流出的現時義務。這裡的義務包括法定義務或推定義務。

A liability is defined as an obligation to transfer economic benefits as a result of a past transaction or event. The obligation may be a legal obligation or a constructive obligation.

A good example of a constructive obligation is a retail store which allows customers to return unwanted goods, a policy which is in excess of the minimum legal obligations to only accept goods which are returned because they faulty or defective in some way.

Unit 1 Provision

1.1 Definition

準備是償付時間和金額不確定的負債。

A provision can be defined as a liability of uncertain timing or amount.

For example, a company may face a legal action for a breach of health and safety law. The likely repercussion is that it may be fined if the court judgement is made against them. The timing and severity of the fine will be decided by a court at some future date. The key question is: should the company reflect this information in its financial statements?

If we assume that the potential fine could be significant (it could even lead to the closure of the business), should the potential consequences be disclosed to the shareholders in some way? As owners of the business, they are entitled to know about this potentially significant issue that could damage the profits and reputation of the business and therefore the financial value of their investment in the company.

1.2 Accounting for a Provision

The first potential course of action management can take is to recognize a provision in the accounts. This is done by estimating the potential cost of the uncertain event and recognizing it immediately. As the amount would be settled at a future date, a corresponding liability is recorded, as follows:

Dr   Expenses
Cr   Provision

準備可以歸類為一項流動負債或非流動負債。

The provision will need to be classified as either a current or non-current liability as be fits the situation.

1.3 Criteria for Recognizing a Provision

Given the uncertainty relating to provisions there is significant scope for accounting error, or even deliberate manipulation of provisions to alter profits.

To reduce this risk IAS 37 Provisions, contingent liabilities and contingent assets provides three criteria that must be met before a provision can be recognized in the financial statements:

· 過去的交易或事項形成的現時義務。

· There must be a present obligation (legal or constructive) that exists as the result of a past even.

· 經濟利益很可能流出企業。

· There must be a probable transfer of economic benefits.

· 金額能夠可靠地計量。

· There must be a reliable estimate of the potential cost.

1.4 Movement in Provision

Provision should be reviewed at each statement of financial position date and adjusted to reflect the current best estimate.

Increase in provision:

Dr   Relevant expense account
Cr   Provision

Decrease in provision:

Dr   Provision
Cr   Relevant expense account

Unit 2   Contingencies

2.1   Contingent Liabilities

或有負債是指：
·過去的交易事項形成的潛在義務。
·過去的交易事項形成的很可能承擔的義務，但是義務金額不能可靠計量的。

A contingent liabilities is defined as:
· A possible obligation that arises from past events.

· A probable obligation that arises from past evens but the amount of the obligation cannot be measured with sufficient reliability.

A contingent liabilities is not recorded in the financial statement. It should be disclosed as a supporting note to the financial statement.

當一項預計負債不能滿足 IAS 37 號準則的確認條件，則不能在財務報告中確認，但應在財務報告中做為或有負債予以披露。

When a provision is not recognized in the financial statements because it does not meet the criteria specified in IAS 37, it may still need to be disclosed as a contingent liability in the financial statements.

Examples of contingent liabilities include outstanding litigation where the potential costs cannot be estimated with any degree of reliability or when the likelihood of losing the litigation is only deemed possible (rather than probable).

2.2   Contingent Assets

或有資產，指過去的事項形成的潛在資產，其存在須通過企業無法完全控制的未來不確定事項的發生或不發生予以證實。

A contingent asset is defined as a possible asset that arises from past events and whose existence will be confirmed only by the occurrence or non-occurrence of one or more uncertain future events not wholly within the control of the enterprise.

An example of a contingent asset is when a business claims compensation from other party and the outcome of the claim is uncertain at the reporting date.

2.3   Accounting for Contingent Liabilities and Contingent Assets

The requirements of IAS 37 regarding contingent liabilities and contingent assets are summarized in the Table 11.1.

Table 11.1

Probability of occurrence	Contingent liabilities	Contingent assets
Virtually certain	Provide	Recognized
Probable	Provide	Disclosure note in financial statements
Possible	Disclosure not in financial statements	Ignore
remote	Ignore	Ignore

Note that the reporting standard gives no guidance regarding the meaning of the terms in the left-hand column. One possible interpretation is as follows:

基本確定 >95%	Virtually certain > 95%
很可能 51%~95%	Probable 51%~95%
可能 5%~50%	Possible 5%~50%
極小可能 <5%	Remote < 5%

Chapter 12 Bank Reconciliation

銀行存款餘額調節表調整：

・銀行存款日記帳餘額；

・銀行對帳單餘額。

The objective of a bank reconciliation is to reconcile the difference between:

・the cash book balance (i.e., the business' record of their bank account);

・the bank statement balance (i.e., the bank's record of the bank account).

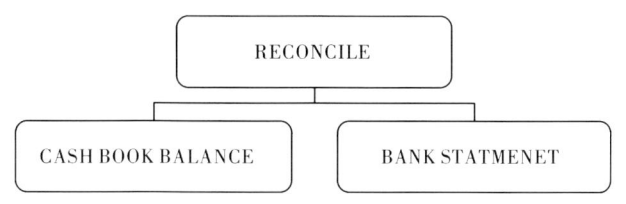

```
            RECONCILE
           /         \
  CASH BOOK BALANCE   BANK STATMENET
```

Unit 1 Bank Statement and Cash Book

1.1 Bank Statement

銀行對帳單

A bank statement is sent by a bank to its short-term receivables and payables – customers with bank overdrafts and customers with money in their account – itemizing the balance on the account at the beginning of the period, receipts into the account, payments from the account during the period, and the balance at the end of the period.

Note:

如果企業有 $6,000 的銀行存款，企業的銀行存款日記帳上為借方餘額 $6,000，但在銀行對帳單上則為貸方餘額 $6,000。

It is necessary to remember, however, if a customer has money in his account, the bank owes him that money, and the customer is therefore a payable of the bank (hence the phrase 『to be in credit』 means to have money in your account). This means that if a business has $6,000 cash in the bank, it will have a debit balance in its own cash book, but for the bank statement, if it reconciles exactly with the cash book, it states that there is a credit balance of $6,000. The bank's records are a 『mirror image』 of the customer's own records, with debits and credits reversed.

1.2 Cash Book

The cash book of a business is the record of how much cash the business believes that it has in the bank. The balance shown by the bank statement should be the same as the cash book balance on the same date. However, you will probably agree that the company's bank statement balance is rarely exactly the same as their own figure.

銀行存款日記帳和銀行對帳單核對不符的原因有：

・差錯

・漏記

・未達帳項

Differences between the cash book and the bank statement arise for three reasons:

· Errors – usually in the cash book.

· Omissions – such as bank charges not posted in the cash book.

· Timing differences – such as unpresented cheques.

1.3 Why is a Bank Reconciliation Necessary

A bank reconciliation is needed to identify and account for the differences between the cash book and the bank statement.

Unit 2 Bank Reconciliation

2.1 Definition

A bank reconciliation is a comparison of a bank statement (sent monthly, weekly or even daily by the bank) with the cash book. Differences between the balance on the bank statement and the balance in the cash book will be errors or timing differences, and they should be identified and satisfactorily explained.

2.2 What to Look for When Doing the a Bank Reconciliation

The cash book and bank statement will rarely agree at a given date. If you are doing a bank reconciliation, you may have to look for the following items.

（a）更正和調整銀行存款日記帳。

(a) Corrections and adjustments to the cash book.

・銀行定期收入或支付的款項，企業尚未登記入帳。	・Payments made into the bank account or from the bank account by way of standing order, which have not yet been entered in the cash book.
・直接存入銀行帳戶的股利或從銀行存款直接支付的股利，企業尚未登記入帳。	・Dividend received, paid direct into the bank account but not yet entered in the cash book.
・企業尚未登記的銀行利息和手續費。	・Bank interest and bank charges, not yet entered in the cash book.
・銀行存款日記帳中需要更正的差錯。	・Errors in the cash book that need to be corrected.
(b) 調整銀行對帳單。	(b) Items reconciling the corrected cash book balance to the bank statement.
・企業開出的支票並記入銀行存款日記帳的貸方，但是銀行尚未辦理轉帳。（未兌付的支票）	・Cheques drawn (i.e., paid) by the business and credited in the cash book, which have not been presented to the bank, or 『cleared』, and so do not yet appear on the bank statement. These are commonly known as unpresented cheques or outstanding cheques.
・企業收到支票，存入銀行，借記銀行存款日記帳，但銀行尚未記帳。（在途存款）	・Cheques received by the business, paid into the bank and debited in the cash book, but which have not yet been cleared and entered in the account by the bank, and so do not yet appear on the bank statement. These are commonly known as outstanding lodgements or deposits credited after date.

Example 12.1

At 31 January 21×6, the balance in the cash book of Wordsworth Co. was $805.15 debit. A bank statement on 30 January 21×6 showed Wordsworth Co. to be in credit by $1,112.30.

On investigation of the difference between the two sums, it was established that:

・The cash book had been undercast by $90.00 on the debit side.

・Cheques paid in not yet credited by the bank amounted to $208.20, called outstanding lodgements.

・Cheques drawn not yet presented to the bank amounted to $425.35, called unpresented cheques.

Required:

(a) Show the correction to the cash book.

(b) Prepare a statement reconciling the balance per bank statement to the balance per cash book.

Solution

(a) Corrected Cash Book

	$
Cash book balance brought forward	805.15
Add	
Correction of undercast	90.00
Corrected balance	895.15

(b) Bank Reconciliation

	$
Balance per bank statement	1,112.30
Add	
Outstanding lodgements	208.20
	1,320.50
Less	
Unpresented cheques	(425.35)
Balance per cash book	895.15

Chapter 13　Correction of Errors

Errors often occur in accounting. Trial balance may help in identifying errors. However not all errors are identifiable by trial balance.

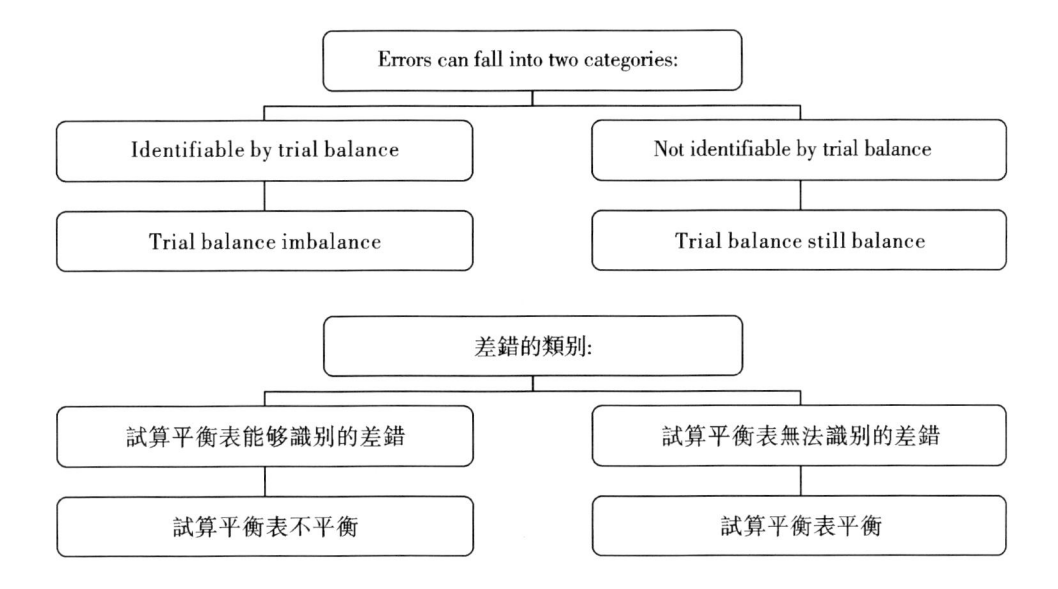

```
                    Errors can fall into two categories:
                   /                                    \
   Identifiable by trial balance              Not identifiable by trial balance
          |                                               |
    Trial balance imbalance                     Trial balance still balance
```

```
                            差錯的類別:
                   /                              \
   試算平衡表能夠識別的差錯              試算平衡表無法識別的差錯
          |                                       |
    試算平衡表不平衡                         試算平衡表平衡
```

Unit 1　Errors Identifiable in Trial Balance

如果試算平衡表的借方欄和貸方欄不相等, 則說明會計記錄有差錯。

(a) 分類帳戶結帳時發生的差錯。

If the two columns of the trial balance are not equal, there must be an error in recoding the transactions in the accounts.

(a) Mistake in balancing off the ledger accounts.

Bank

Capital	$10,000	Electricity	$3,000
Sales	$5,000	Bal c/f	$14,000　(supposed to be $12,000)
	$15,000		$15,000
Bal b/f	$14,000		

Capital

		Bank	$ 10,000

Sales

		Bank	$ 5,000

Electricity

Bank	$ 3,000		

Trial Balance

	Dr	Cr
	$	$
Bank	14,000	
Capital		10,000
Sales		5,000
Electricity	3,000	
	17,000	15,000
Diff		2,000

（b）編製試算平衡表時發生的差錯。

(b) Mistake in drafting the trial balance. For example, amount was wrongly transferred to trial balance.

Bank

Capital	$10,000	Electricity	$3,000
Sales	$5,000	Bal c/f	$12,000
	$15,000		$15,000
Bal b/f	($12,000)		

Capital

		Bank	$ 10,000

Sales

		Bank	$ 5,000

Electricity

Bank	$ 3,000		

Trial Balance

	Dr	Cr
	$	$
Bank	11,000	
Capital		10,000
Sales		5,000
Electricity	3,000	
	14,000	15,000
Diff		(1,000)

(c) 餘額遺漏。

(c) Omission. For example, Electricity balance was not transferred to the trial balance.

Bank

Capital	$ 10,000	Electricity	$ 3,000
Sales	$ 5,000	Bal c/f	$ 12,000
	$ 15,000		$ 15,000
Bal b/f	$ 12,000		

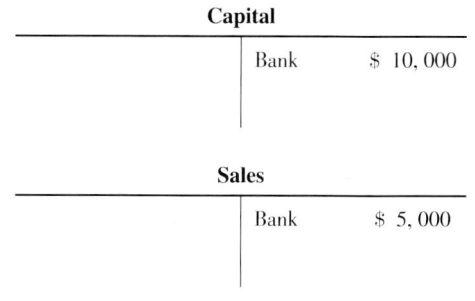

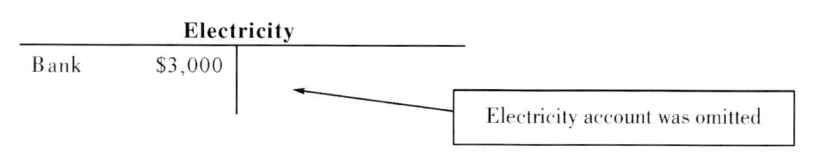

Capital

		Bank	$ 10,000

Sales

		Bank	$ 5,000

Electricity

Bank	$3,000		

← Electricity account was omitted

Trial Balance

	Dr	Cr
	$	$
Bank	12,000	
Capital		10,000
Sales		5,000
	12,000	15,000
Diff		(3,000)

(d) 遺漏某一個方向的會計記錄。

(d) Single entry error. Either the debit being posted or credit is being posted.

Credit Sales

Dr	Trade accounts receivable	$ 2,000
Cr	Sales	$ 2,000

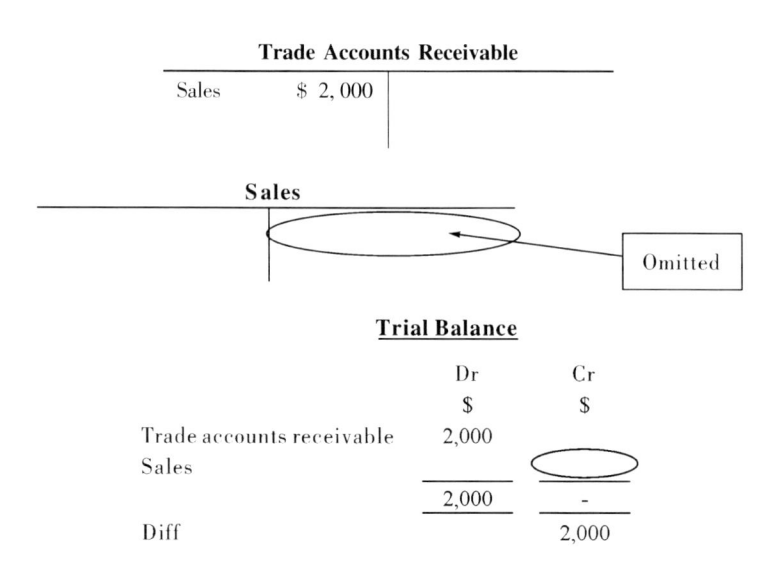

Trade Accounts Receivable

Sales	$ 2,000	

Sales

Omitted

Trial Balance

	Dr	Cr
	$	$
Trade accounts receivable	2,000	
Sales		
	2,000	-
Diff		2,000

(e) 過帳時借貸方金額不相等。

(e) Debit or credit are posted with different amount.

Cash Sales

Dr	Bank	$ 1,570
Cr	Sales	$ 1,570

Bank

Sales	$1,570	

Sales

	Bank	$1,600

Trial Balance

	Dr	Cr
	$	$
Bank	1,570	
Sales		1,600
	1,570	1,600
Diff		(30)

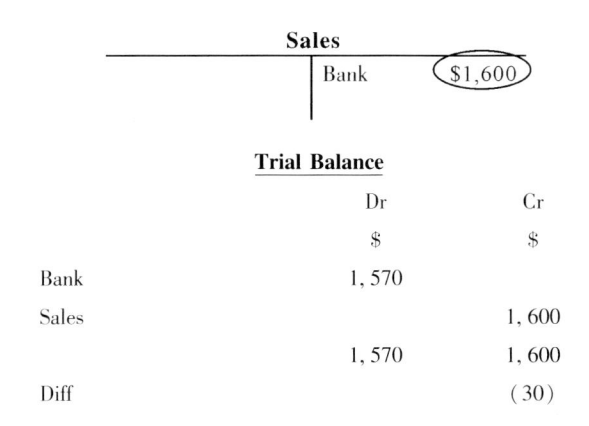

(f) 借方或貸方金額過帳時記錄了兩次。

(f) Either Debit is posted twice or credit is posted twice.

Cash Sales

Dr	Bank	$ 1,600
Cr	Sales	$ 1,600

Bank

Sales	$ 1,600	
Sales	$ 1,600	
	$ 3,200	

Sales

	Bank	$ 1,600

Trial Balance

	Dr	Cr
	$	$
Bank	3,200	
Sales	———	1,600
	3,200	1,600
Diff		1,600

(g) 借方或貸方數字錯位。

(g) Single transposition error (2 digits are recorded in wrong order, on one side of the entry only).

Cash Sales

Dr	Bank	$ 1,360
Cr	Sales	$ 1,360

Bank

Sales	$ 1,360	

Sales

	Bank	$1,630

Trial Balance

	Dr	Cr
	$	$
Bank	1,360	
Sales		1,630
	1,360	1,630
Diff		(270)

如果借貸方的差額可以被 9 整除，則可能出現數字錯位的差錯。上例中，$ 270 ÷ 9 = $ 30。

You can often detect a transposition error by checking whether the difference between debits and credits can divided exactly by 9. In the above example, $ 270 ÷ 9 = $ 30.

Unit 2 Suspense Account

開設臨時帳戶有以下兩個原因：

・試算平衡表不平衡，借貸方的差額計入「臨時帳戶」。

・簿記員運用復式記帳系統登記經濟業務，不能確定某一方向的會計記錄時，可以暫時記入臨時帳戶。

There are two main reason why suspense accounts may be created：

・When the trial balance is imbalance, the difference need to be recorded in a temporary account called 『suspense account』.

・When a bookkeeper performing double entry is not sure where to post one side of an entry they may debit or credit a suspense account and leave the entry until its ultimate destination is clarified.

Example 13.1

(a) A cash sale of $500 has been transacted, only cash account was debited with $500, and the credit entry was omitted.

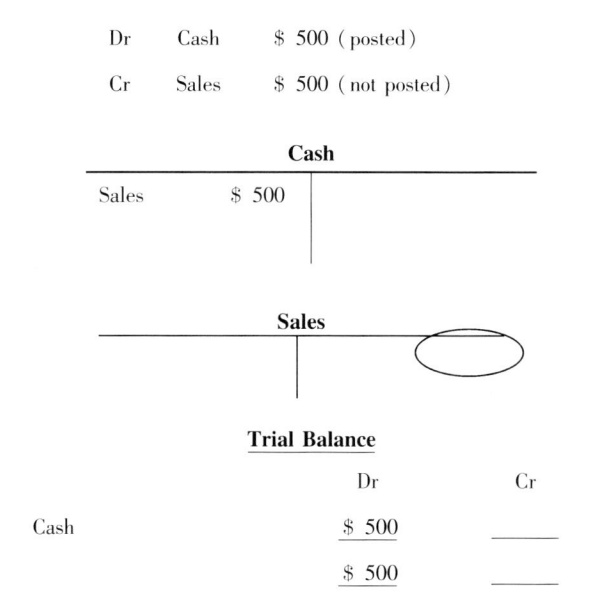

    Dr   Cash $500 (posted)
    Cr   Sales $500 (not posted)

Cash

Sales	$500	

Sales

	◯

Trial Balance

	Dr	Cr
Cash	$500	
	$500	

開設「臨時帳戶」，記錄借貸方差異額。

(b) A suspense account is opened to temporary record the difference.

Trial Balance

	Dr	Cr
	$	$
Cash	500	
Suspense account		500
	500	500

Cash

Sales	$500	

Suspense a/c

	$500

現金帳戶的記錄是正確的，因此差錯更正如下：

Since cash has been correctly debited, therefore are correction will be as follow:

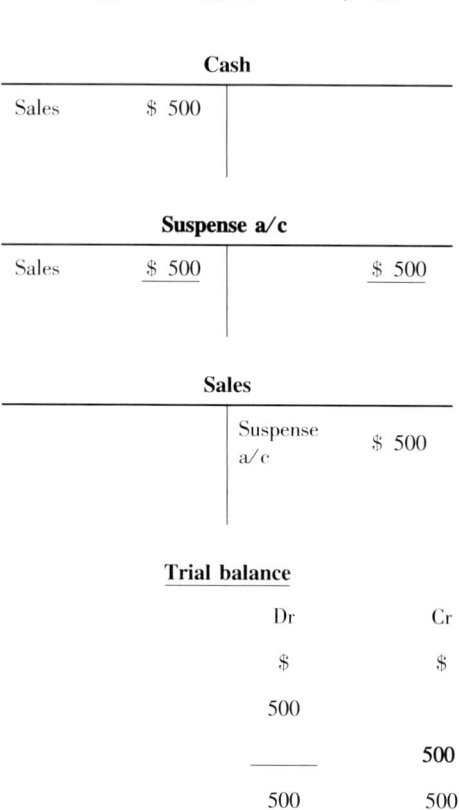

Dr　　Suspense　　$ 500（transfer out the error）
Cr　　Sales　　　　$ 500

Cash

Sales	$ 500		

Suspense a/c

Sales	$ 500		$ 500

Sales

		Suspense a/c	$ 500

Trial balance

	Dr	Cr
	$	$
Cash	500	
Sales		500
	500	500

臨時帳戶是臨時開設的。在會計期末編製財務報表時臨時帳戶應當全部註銷。

Suspense accounts are only temporary. None should exist when it comes to drawing up the financial statement at the end of the accounting period.

Unit 3　Error Not Identifiable in Trial Balance

試算平衡表不能查找出下列類型的差錯。

（a）遺漏了整個經濟業務的會計記錄。

A trial balance, however, will not disclose the following types of errors.

（a）Errors with omissions, an entry is totally omitted.

Cash Sales

Dr	Bank	$ 1,560
Cr	Sales	$ 1,560
Dr	Bank	$2,150 ← Omitted
Cr	Sales	$2,150

Bank

Sales $ 1,560	

Sales

	Bank $ 1,560

Trial Balance

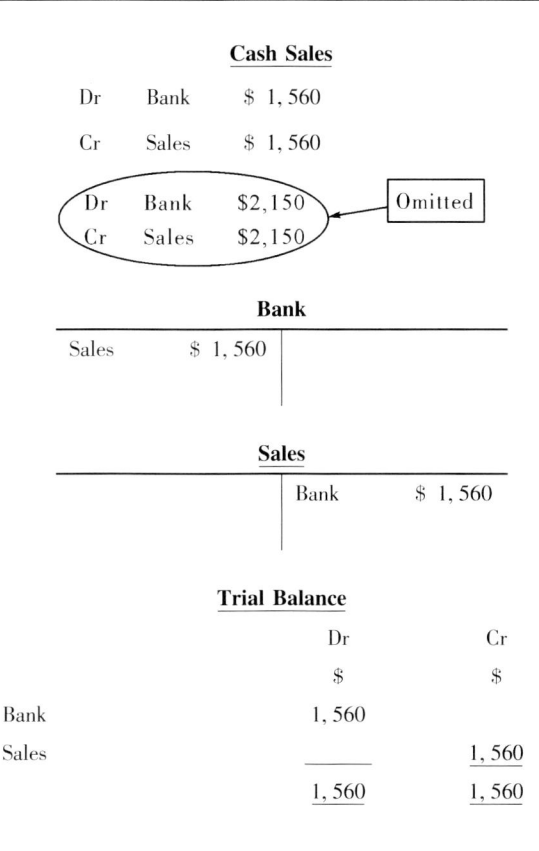

	Dr	Cr
	$	$
Bank	1,560	
Sales		1,560
	1,560	1,560

The correcting entries required as follows:

Dr	Bank	$ 2,150
Cr	Sales	$ 2,150

(b) 金額被過入同一類型的其他錯誤帳戶。

例如:

・租金費用被錯誤地記成電費，但是兩個都是費用類帳戶。

・機械設備被錯誤地記成機動車輛，但兩個都是非流動資產帳戶。

(b) Error of commission – amount was posted to an wrong account of same class.

For example:

・Rental account was recorded in electricity account, but both are expense accounts.

・Machinery was recorded to motor vehicles, but both are non-current asset accounts.

Expense payment

```
Dr    Electricity    $500      ← Rental expense
Cr    Bank           $500
```

Bank

	$ 500

Electricity

$ 500	

Trial Balance

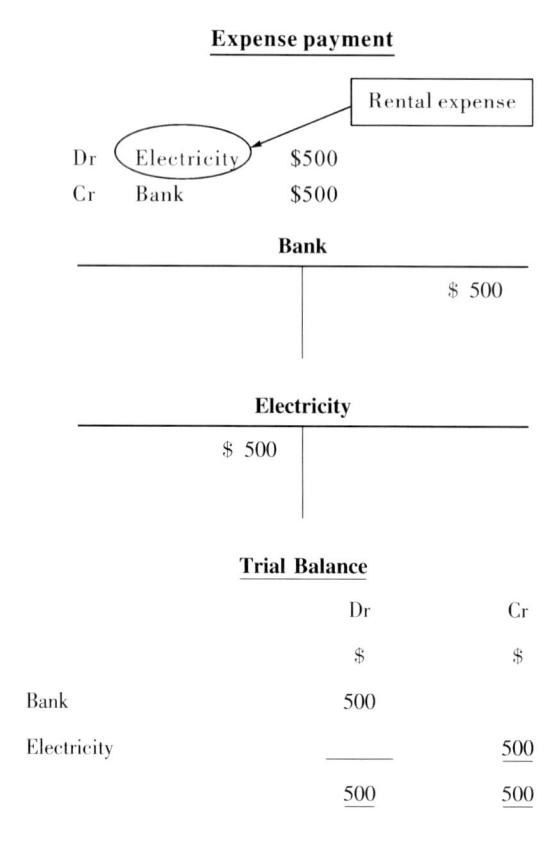

	Dr	Cr
	$	$
Bank	500	
Electricity		500
	500	500

The correcting entries required as follows:

```
Dr    Rental        $ 500
Cr    Electricity   $ 500
```

(c) 借方或貸方被過入非同一類型的錯誤帳戶。

例如，汽車修理費用（費用，利潤表項目）被過入機動車輛帳戶（資產、資產負債表項目）。

(c) Error of principal – debit or credit is posted into wrong account of difference class.

For example, motor repair expense (expenses, go to statement of profit or loss) was posted into motor vehicle account (asset, go to statement of financial position).

Expense payment

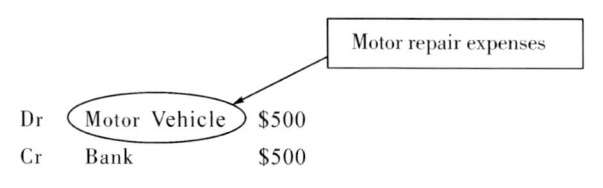

```
Dr    Motor Vehicle    $500    ← Motor repair expenses
Cr    Bank             $500
```

```
                        Bank
─────────────────────────────────────────
                        │  $ 500
                        │
                        │
                  (Motor vehicle)
─────────────────────────────────────────
           $500         │
                        │
```

                        Trial Balance

	Dr	Cr
	$	$
Bank	500	
Motor Vehicle		500
	500	500

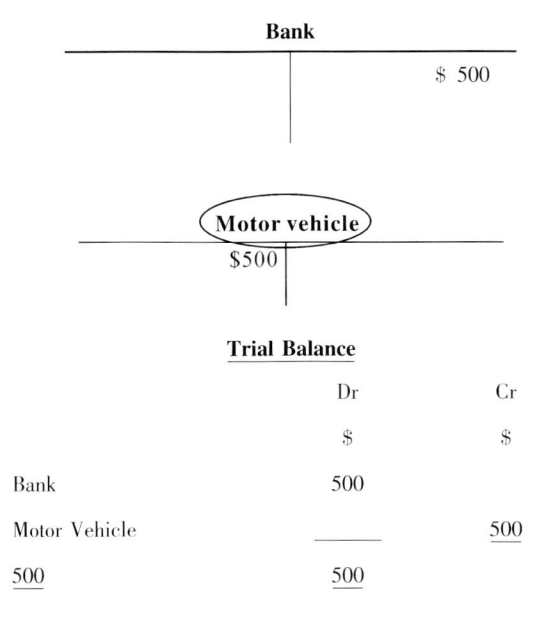

The correcting entries required as follows:

    Dr Motor Repair expenses $ 500
    Cr   Motor Vehicle $ 500

(d) 借貸金額錯誤。 (d) Error of original entry. Both debit and credit are posted with wrong amount.

      Cash Sales
  Dr Bank $ 1,930
  Cr Sales $ 1,930

```
                       Bank
─────────────────────────────────────────
   Sales    ($1,390)   │
                       │

                       Sales
─────────────────────────────────────────
                       │  Bank    ($1,390)
                       │
```

Trial Balance

	Dr	Cr
	$	$
Bank	1,390	
Sales		1,390
	1,390	1,390

The correcting entries required as follows:

Dr	Bank	$ 540
Cr	Sales	$ 540

（e）相互抵消的差錯。　　（e）Compensating error. Errors that cancel out each other. For example. Two transposition errors of $540 might occur in extracting ledger balances, one on each side of the double entry.

Dr	Administration expense	$2,282
Cr	Bank	$2,282

Dr	Bank	$8,391
Cr	Sundry income	$8,391

Bank

$ 8,391	$ 2,282

Administration Expense

Bank	$2,822		

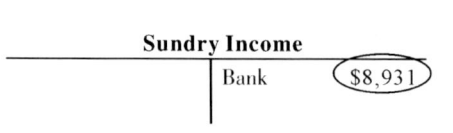

Sundry Income

		Bank	$8,931

Trial Balance

	Dr	Cr
	$	$
Bank	6,109	
Administration expense	2,822	
Sundry income		8,931
	8,931	8,931

The correcting entries required as follows:

Dr	Sundry income		$ 540
Cr	Administration expense		$ 540

Chapter 14 Preparing a Trial Balance

試算平衡表是一張包含所有總分類帳戶的列表。

A trial balance is a list of all the general ledger accounts (both revenue and capital) contained in the ledgers of a business. This list will contain the name of each nominal ledger account and the value of that nominal ledger balance. Each nominal ledger account will hold either a debit balance or a credit balance. The debit balance values will be listed in the debit column of the trial balance and the credit value balance will be listed in the credit column. The trading profit and loss statement and the statement of financial position and other financial reports can then be produced using the ledger accounts listed on the trial balance.

借方餘額列示在試算平衡表的借方欄，貸方額列示在試算平衡表的貸方欄。

Unit 1 Trial Balance

在分類帳戶中，所有的經濟業務都被記入一個帳戶的借方和另外一個帳戶的貸方，其借方貸方金額相等。如果借貸金額不相等，則帳戶記錄有差錯。

Within the general ledger, all transactions are recorded as a debit entry in one account and a credit entry in another account, and the total value of debit entries and credit entries must always be the same. If they are not, something has gone wrong. Trial balance is a long list of all the ledger balances created for a specific date showing the respective debit and credit balances. It is used to check whether the ledger accounts are correct, in so far as the total debit balances and total credit balances are equal.

試算平衡表用於檢驗帳戶記錄是否正確。

1.1 The Reason for a Trial Balance

There are three main reasons for preparing a trial balance:

・在會計期末，編製利潤表和資產負債表前，先編製試算平衡表。

・At the end of the financial year, a trial balance is used as a starting point for preparing a statement of profit or loss for the year and a statement of financial position as at the year end.

・試算平衡表列示了所有資產、負債、所有者權益、收入、費用帳戶的餘額。

・編製試算平衡可以查找出帳戶記錄中的某些類型的差錯。如果試算平衡表借貸不相等，則帳戶記錄中有差錯。

・A trial balance shows the current balances on all the asset, liability, capital, income and expense accounts. This can provide useful information to management.

・In a manual accounting system, preparing a trial balance is a procedure for identifying certain types of errors in the accounts. If the total of debit balances and the total of credit balances are not equal, there must have been a mistake (or several mistakes) in entering transactions in the ledger accounts. Once the existence of an error has been identified, the next step is to carry out an investigation, and try to find where the double entry mistaken or mistakes have happened. Once the error has been found, it should be corrected.

1.2 Producing a Trial Balance

There are two stages in preparing a trial balance.

步驟一，結出各帳戶的餘額。

Step 1, balance off all the accounts in the general ledger.

步驟二，將餘額填入試算平衡表，加總借方餘額合計和貸方餘額合計。

Step 2, list all the accounts in the ledger, with their debit or credit balance, and add up the total debit balances and the total credit balances.

At the year end, the ledger accounts must be closed off in preparation for the recording of transactions in the next accounting period.

(1) Statement of financial position ledger accounts

結出的帳戶餘額是該帳戶當期的期末餘額和下一期的期初餘額。

Balancing the account will result in a balance c/f (being the asset/liability/capital at the end of the accounting period) and a balance b/d (being the asset/liability/capital at the start of the next accounting period).

Example 14.1

Balance of the following account:

Bank

	$		$
Capital	10,000	Purchase	200
Sales	250	Rent	150
		Electricity	75
		Balance c/d	9,825
	10,250		10,250
Balance b/d	9,825		

(2) Profit or loss ledger accounts

會計期末，收入和費用帳戶當期的發生額轉入「利潤或損失」帳戶。

At the end of a period any amounts that relate to that period are transferred out of the income and expenditure accounts into another ledger account called 『profit or loss』.

Example 14.2

Balance of the following account:

Sales

	$		$
Return	500	Bank	1,000
Taken to P/L	2,000	Receivables	1,500
	2,500		2,500

Profit or loss account

	$		$
		Sales	2,000

Unit 2 Steps of Preparing a Trial Balance

At the end of the year, once all ledger accounts have been balanced off, the closing balances are summarised on a long list of balances. This is referred to as a trial balance. All the closing debit balances are summarised in one column and the closing credit balances in another. Given the nature of the double entry system described in this text the totals of both columns should be agree. If not, the discrepancy must be investigated and corrected. The layout of a trial balance is illustrated below:

依據復式記帳系統的記帳規則，試算平衡表的借貸合計相等。如果不相等，應查找並更正錯誤。

Table 14.1　　Trial Balance as at 31 December 20××

	Dr	Cr
	$	$
Purchases	×	
Non-current assets	×	
Trade receivables	×	
Cash	×	
Share capital		×
Loans		×
Trade payable		×
Profit or loss account		×
	×	×

Example 14.3

Below are the ledger accounts of Lucas as at 31 December 20×7. We should balance the accounts, bring down the balances and show all the balances in a trial balance.

Step 1: balance each account and bring down the balances.

Bank

	$		$
Capital	10,000	Purchase	305
Sales	300	Motor car	400
Receivables	100	Payables	200
Loan	600	Rent	40
		Balance c/d	1,055
	2,000		2,000
Balance b/d	1,055		

Capital

	$		$
Balance c/d	1,000	Bank	1,000
	1,000		1,000
		Balance b/d	1,000

Motor Car

	$		$
Bank	400	Balance c/d	400
	400		400
Balance b/d	400		

Payables

	$		$
Bank	200	Purchases	400
Balance c/d	200		
	400		400
		Balance b/d	200

Receivables

	$		$
100	Sales	250	Bank	150
		Balance c/d	150
	250		250
Balance b/d	150		

Loan

	$		$
Balance c/d	600	Bank	600
	600		600
		Balance b/d	600

Purchases

	$		$
Bank	305		
Payables	400	Balance c/d	705
	705		705
Balance b/d	705		

Sales

	$		$
		Bank	300
P/L account	550	Receivables	250
	550		550

Rent

	$		$
Bank	40	P/L account	40
	40		40

Profit or loss account

	$		$
Rent	40	Sales	550

注意「利潤或損失」帳戶至此尚未結出餘額。該帳戶的餘額將轉入「留存收益」帳戶。

Note that the 『profit or loss』 account is not closed off at this stage. The balance of this account will be transferred to the 『retained earnings』 account in the adjustment process which will be described in a later chapter.

Step 2: Prepare the trial balance showing each of the balances in the ledger accounts.

Table 14.2

Lucas
Trial Balance as at 31 December 20××

	Debit	Credit
	$	$
Bank	1,055	
Capital		1,000
Motor car	400	
Payables		200
Receivables	150	
Loan		600
Purchases	705	
Profit or loss	40	550
	2,350	2,350

The total debit balances and the total credit balances are the same, $ 2,350. It therefore appears that the double entry accounting has been consistently carried out.

Chapter 15   Prepare Basic Financial Statements

會計報表，也叫財務報表。

Accounting statements are the end product of the accounting process, communicating important accounting information to users. The accounting statements, also called financial statements, are the means conveying to the management and the interested outsider a concise picture of the profitability and financial position of a business. The basic financial statements include the statement of financial position, statement of profit or loss and the statement of cash flows.

財務報表包括資產負債表、利潤表和現金流量表。

Unit 1   Statement of Financial Position

資產負債表示反應企業某一特定時點的負債、所有者權益和資產的報表。

The statement of financial position is a statement of the liabilities, equity and assets of a business at a given moment in time. It is like a 『snapshot』 photograph, since it captures on paper a still image, frozen at a single moment in time, of something which is dynamic and continually changing. Typically, a statement of financial position is prepared at the end of accounting period to which the financial statements relate.

1.1   Format of a Statement of Financial Position

資產負債表與會計等式的差異有如下兩點：

A statement of financial position is very similar to the accounting equation. In fact, there are only two differences between a statement of financial position and accounting equation as follows.

·負債和資產列示的方法和格式。

· The manner or format in which the liabilities and assets are presented.

·資產負債表包含更詳細的信息。

· The extra detail which is usually contained in a statement of financial position.

Table 15.1　　　　　　　**Statement of Financial Position as at 31 Dec 20××**

	$	$
Non-current assets		
Property	×	
Equipments	×	
		×
Current assets		
Inventory	×	
Trade receivables	×	
Bank	×	
		×
		×
Capital		
Opening balance	×	
Profit for the year	×	
Less: Drawings	(×)	
		×
Non-current liabilities		×
Current liabilities		
Trade payables	×	
Accruals	×	
		×
		×

Note that the suggested statement of financial position format makes a distinction between current and non-current assets and liabilities.

1.2　Current Assets

流動資產：

・為交易目的而持有；

・預計在 12 個月內變現；

・現金或現金等價物。

主要的流動資產包括存貨、應收帳款、現金等。

An asset should be classified as a current asset if it is:

・held primarily for trading purposes；

・expected to be realised within 12 months of the statement of financial position date；

・cash or a cash equivalent (i.e., a short time investment, such as a 30 day bond).

The main items of current assets are therefore include inventories, receivables, cash, etc.

1.3　Non-current Assets

A non-current asset is an asset acquired for continuing use within the business, with a view to earning income or making profits from its use, either directly or indirectly. It is not acquired for sale to a customer.

製造業中，生產設備是一項非流動資產。

服務業中，員工向客戶提供勞務時使用的生產設備也屬於非流動資產。

廠房、辦公家具、計算機、汽車、貨車、倉庫用的托盤都是非流動資產。

In a manufacturing industry, a production machine is a non-current asset, because it makes goods which are then sold.

In a service industry, equipment used by employees giving service to customers is a non-current asset (e.g., the equipment used in a garage, and furniture in a hotel).

Less obviously, factory premises, office furniture, computer equipment, company cars, delivery vans or pallets in a warehouse are all non-current assets.

1.4　Current Liabilities

流動負債：

· 預計在一個正常的營業週期內清償。

· 為交易目的而持有。

· 資產負債表日起12個月內到期應予以清償。

· 企業無權將清償推遲至資產負債表日後12個月以上的負債。

A liability should be classified as a current liability if:

· It is expected to be settled in the normal course of the enterprise's operating cycle.

· It is held primarily for the purpose of being traded.

· It is due to be settled within 12 months of the statement of financial position date or.

· The company does not have an unconditional right to defer settlement for at least 12 months after the statement of financial position date.

Examples of current liabilities are:

· 一年內償付的借款；

· 銀行透支；

· 應付帳款；

· 應交稅費。

· loans repayable within one year;

· a bank overdraft which is usually repayable on demand;

· trade accounts payable;

· tax payable。

1.5　Non-current Liabilities

非流動負債是指短期內不會償付的債務，因此流動負債以外的負債就是非流動負債。

A non-current is a debt which is not payable within the 『short term』(i.e., it will not be liquidated shortly) and so any liability which is not current must be non-current.

Examples of non-current liabilities are as follows:

· 償還期在一年以上的借款。

· Loans which are not repayable for more than one year, such as a bank loan or a loan from an individual to a business.

・ A mortgage loan, which is a loan specifically secured against a property. (If the business fails to repay the loan, the lender then has 『first claim』 on the property and is entitled to repayment from the proceeds of the enforced sale of the property.)

・ Loan stock. These are common with limited liability companies. Loan stocks are securities issued by a company at a fixed rate of interest. They are repayable on agreed terms by a specified date in the future. Holders of loan stocks are therefore lenders of money to a company. Their interests, including security for the loan, are protected by the terms of a trust deed.

Unit 2 Statement of Profit or Loss

The statement of profit or loss is a statement in which revenues and expenditure are matched to arrive at a figure of profit or loss. It shows in detail how the profit (or loss) of a period has arisen.

2.1 Format of a Statement of Profit or Loss

Many businesses try to distinguish between a gross profit earned on trading, and a net profit after other income and expenses. In the first part of the statement, revenue from selling goods is compared with direct costs of acquiring or producing the goods sold to arrive at a gross profit figure. From this, deductions are made in the second half of the statement (which we will call the income and expenses section) in respect of indirect costs and additions for non-trading income.

2.2 Gross Profit

The first part shows the gross profit for the accounting period. Gross profit is the difference between (a) and (b) below.

(a) The value of sales.

(b) The purchase costor production cost of the goods sold.

$$\text{Gross profit} = \text{Sales} - \text{Cost of sales}$$

商品的生產成本包括完工產品耗用的原材料的成本，加工成本和生產過程的間接成本。

In the retail business, the cost of the goods sold is their purchase cost from the suppliers. In a manufacturing business, the production cost of goods sold is the cost of raw materials in the finished goods, plus the cost of the labour required to make the goods, and often plus an amount of production 『overhead』 costs.

2.3 Net Profit

淨利潤包括毛利加上銷售收入以外的其他收入減去銷售成本以外的其他費用。

The profit or loss account shows the net profit of the business. The net profit is that the gross profit plus any other income from sources other than the sales of goods minus other expenses of the business, not included in the cost of goods sold.

$$\text{Net profit} = \text{GP} + \text{Non-trading income} - \text{Expenses}$$

A sample of the statement of profit or loss is shown as Table 15.2.

Table 15.2　　Statement of Profit or Loss for the Year End 31 Dec 20××

	$	$
Revenue		×
Cost of sales		
Opening inventory	×	
Purchases	×	
	×	
Less: closing inventory	×	
		(×)
Gross profit		×
Other income		×
Expenses		
Electricity	×	
Telephone expenses	×	
Allowance for receivables	×	
		(×)
Profit for the year		×

Unit 3 Preparing Financial Statements

The financial affairs of Newstart Tools prior to the commencement of trading were as Table 15.3 shows.

Newstart Tools
Table 15.3   Statement of Financial Position at as 1 August 20×5

	$	$
Non-current assets		
Motor vehicle		2,000
Shop fittings		3,000
		5,000
Current assets		
Inventories		12,000
Cash		1,000
Capital		12,000
Current liabilities		
Bank overdraft	2,000	
Trade payables	4,000	
		6,000
		18,000

At the end of six months the business had made the following transactions.

(a) 賒購商品, 目錄價格 $ 10,000。

(b) 商業折扣為目錄價格的 2%。支付供應商部分貨款 $ 8,000, 並得到 5%現金折扣。

(c) 期末存貨價值 $ 5,450。

(d) 賒銷總額為 $ 27,250。

(e) 20×6 年 1 月 31 日應收帳款餘額為 $ 3,250, 其中 $ 250已註銷。壞帳準備計提比例為 2%。

(a) Goods were purchased on credit at a list price of $ 10,000.

(b) Trade discount received was 2% on list price and there was a settlement discount received of 5% on settling debts to suppliers of $ 8,000. These were the only payments to suppliers in the period.

(c) Closing inventories of goods were valued at $ 5,450.

(d) All sales were on credit and amounted to $ 27,250.

(e) Outstanding receivables balance at 31 January 20×6 amounted to $ 3,250 of which $ 250 were to be written off. An allowance for receivables is to be made amounting to 2% of the remaining outstanding receivables.

(f) 支付了各項費用。

(f) Cash payments were made in respect of the following expenses:

	$
Stationery, postage and wrapping	500
Telephone charges	200
Electricity	600
Cleaning and refreshments	150

(g) 所有者提款 $ 6,000。

(g) Cash drawings by the proprietor, amounted to $ 6,000.

(h) 清償20×5年8月1日的銀行透支。支付透支的利息和銀行手續費用 $ 40。

(h) The outstanding overdraft balance as at 1 August 20×5 was paid off. Interest charges and bank charges on the overdraft amounted to $ 40.

Prepare the statement of profit or loss of Newstart Tools for the six months to 31 January 20×6 and a statement of financial position as at that date. Ignore depreciation.

The answer of these transactions are shown as Table 15.4 and Table 15.5.

Table 15.4

Statement of Profit or Loss
For The Six Months Ended 31 January 20×6

	$	$
Revenue		27,250
Cost of sales		
Opening inventory	12,000	
Purchases	9,800	
	21,800	
Closing inventory	5,450	
		16,350
Gross profit		10,900
Other income – discounts received		400
		11,300
Expenses		
Stationery, postage and wrapping	500	
Telephone charges	200	
Electricity	600	
Cleaning and refreshments	150	
Irrecoverable debts written off	250	
Allowance for receivables	60	
Interest and bank charges	40	

Table 15.4 (Continued)

		1,800
Profit for the period		9,500

Table 15.5

Newstart Tools
Statement of Financial Position as at 31 January 20×6

	$	$
Non-current assets		
Motor vehicle	2,000	
Shop fittings	3,000	
		5,000
Current assets		
Inventories	5,450	
Receivables, less allowance for receivables	2,940	
Cash	7,910	
		16,300
		21,300
Capital		12,000
Capital at 1 August 20×5	12,500	
Profit for the period	9,500	
	21,000	
Less drawings	6,000	
		15,500
Current liabilities		
Trade payables		5,800
		21,300

Notes
(a) Purchases at cost $10,000 less 2% trade discount.
(b) 5% of $8,000 = $4,000.
(c) Expenses are grouped into sales and distribution expenses (here assumed to be electricity, stationery and postage, bad debts and allowance for receivables) administration expenses (here assumed to be telephone charges and cleaning) and finance charges.
(d) 2% of $3,000 = $60.
(e) Payables as at 31 January 20×6.
The amount owing to payables in the sum of the amount owing at the beginning of the period, plus the cost of purchases during the period (net of all discounts), less the

payments already made for purchases.

	$
Payables as at 1 August 20×5	4,000
Add purchases during the period, net of trade discount	9,800
	13,800
Less settlement discount received	(400)
	13,400
Less payments to payables during the period*	(7,600)
	5,800

*：$ 8,000 less cash discount of $ 400

(f) Cash at bank and in hand at 31 January 20×6, You need to identify cash payments received and cash payments made.

	$
Cash received from sales	
Total sales in the period	27,250
Add receivables as at 1 August 20×5	0
	27,250
Less unpaid debts as at 31 January 20×6	3,250
Cash received	24,000

國家圖書館出版品預行編目(CIP)資料

財務會計 / 秦戈雯、池昭梅 主編. -- 第一版.
-- 臺北市 : 崧博出版 : 財經錢線文化發行, 2018.10
　　面 ; 　公分
雙語版

ISBN 978-957-735-592-8(平裝)

1. 財務會計

495.4　　　107017193

書　名：財務會計(雙語版)
作　者：秦戈雯、池昭梅 主編
發行人：黃振庭
出版者：崧博出版事業有限公司
發行者：財經錢線文化事業有限公司
E-mail：sonbookservice@gmail.com
粉絲頁　　　　　　網　址：
地　址：台北市中正區延平南路六十一號五樓一室
8F.-815, No.61, Sec. 1, Chongqing S. Rd., Zhongzheng Dist., Taipei City 100, Taiwan (R.O.C.)
電　話：(02)2370-3310　傳　真：(02) 2370-3210
總經銷：紅螞蟻圖書有限公司
地　址：台北市內湖區舊宗路二段 121 巷 19 號
電　話：02-2795-3656　傳真：02-2795-4100　網址：
印　刷：京峯彩色印刷有限公司（京峰數位）

　　本書版權為西南財經大學出版社所有授權崧博出版事業有限公司獨家發行電子書及繁體書繁體版。若有其他相關權利及授權需求請與本公司聯繫。

定價：300元
發行日期：2018 年 10 月第一版
◎ 本書以POD印製發行